W9-CAR-151

David Knight teaches at the University of Durham. He is the author of *Atoms and Elements* and *Natural Science Books in English, 1600–1900.*

THE SOURCES OF HISTORY:
STUDIES IN THE USES OF HISTORICAL EVIDENCE

GENERAL EDITOR: G. R. ELTON

Books in This Series

Volumes already published

G. R. Elton	ENGLAND, 1200–1640
T. H. Hollingsworth	HISTORICAL DEMOGRAPHY
C. L. Mowat	GREAT BRITAIN SINCE 1914
Ian Jack	WALES, 400–1542
Kathleen Hughes	EARLY CHRISTIAN IRELAND
Charles H. Carter	THE WESTERN EUROPEAN POWERS, 1500–1700
Walter Ullmann	LAW AND POLITICS IN THE MIDDLE AGES
W. R. Brock	THE UNITED STATES, 1789–1890
David Knight	SOURCES FOR THE HISTORY OF SCIENCE, 1660–1914
Bruce Webster	SCOTLAND FROM THE ELEVENTH CENTURY TO 1603

The Sources of History:
Studies in the Uses of Historical Evidence

Sources for the History of Science 1660–1914

by

DAVID KNIGHT

CORNELL UNIVERSITY PRESS
Ithaca, New York

First published 1975 by Cornell University Press

International Standard Book Number 0-8014-0941-1
Library of Congress Catalog Card Number 74-19776

Printed in Great Britain

Contents

General Editor's Introduction

By what right do historians claim that their reconstructions of the past are true, or at least on the road to truth? How much of the past can they hope to recover: are there areas that will remain for ever dark, questions that will never receive an answer? These are problems which should and do engage not only the scholar and student but every serious reader of history. In the debates on the nature of history, however, attention commonly concentrates on philosophic doubts about the nature of historical knowledge and explanation, or on the progress that might be made by adopting supposedly new methods of analysis. The disputants hardly ever turn to consider the materials with which historians work and which must always lie at the foundation of their structures. Yet, whatever theories or methods the scholar may embrace, unless he knows his sources and rests upon them he will not deserve the name of historian. The bulk of historical evidence is much larger and more complex than most laymen and some professionals seem to know, and a proper acquaintance with it tends to prove both exhilarating and sobering—exhilarating because it opens the road to unending inquiry, and sobering because it reduces the inspiring theory and the new method to their proper subordinate place in the scheme of things. It is the purpose of this series to bring this fact to notice by showing what we have and how it may be used.

G. R. ELTON

Preface

This book is not a history of science. It is an introduction to the sources for the history of science. History can only be written from the sources we have, and the character of the sources determines what questions we can and cannot profitably ask; if the sources do not enable us to answer a question, there is little point in asking it. Here then we shall give some indication of the questions which are being, or might be, asked in the history of science; and of the answers which are being given.

It will, I hope, become evident that the sources for the history of science do not differ in kind from those used by other kinds of historians; and that the questions and answers raised and given by historians of science are also similar to those which come up in other branches of history. The time when the history of science was dominated by elderly scientists following the progress of theories or experiments important to them, or by philosophers examining the structure of arguments but not much interested in historical situations, has passed. There is still room for those wishing to legitimise a scientific theory, or a view of scientific explanation, by means of carefully-chosen examples from the history of science; in this field as in others, the more the merrier. But such writing on the history of science is no longer, and must not be again, the norm. Professor A. R. Hall, in his Presidential Address to the British Society for the History of Science in 1968, asked: 'Can the History of Science be History?'[1] The answer surely must be 'yes' because if it is not history then it is nothing.

The points made are, I hope, general ones; but many of the examples chosen come from the history of science in Britain. It would be absurd to claim that any one nation has monopolised the development of science, and to study chiefly science in one

[1] *British Journal for the History of Science*, 4 (1969) 207–20.

country is not to subscribe to chauvinism. Science is a cosmopolitan activity; but it is carried on by people living in a particular country at a particular time, and therefore differing in their social organisation, their practical activities, and their modes of thought. Different sciences have flourished in different countries at different times; and part of the historian's task is to find out why. Sources of a similar kind will be found in different countries but we shall not construct from them the same story.

My ignorance has therefore restricted the choice of examples; and similarly where egregious errors or naïve assumptions are described, these were usually committed or held by me unless otherwise ascribed, and are not simply the flogging of dead horses. The notes are intended to refer particularly to works which illuminate or refer to sources rather than to give a fair sample of the literature. The book will be a success if it sends historians or scientists to the sources of the history of science and gives them some idea of fruitful lines of approach; we may hope that their ingenuity will lead to the use of new sources and to the discovery of new information from the old ones.

DURHAM

CHAPTER 1

The History of Science

To the alarm of his contemporaries, George Stephenson carried the Liverpool and Manchester Railway over the swamps of Chatmoss upon 'hurdles thickly interwoven with twisted heath' and covered with a layer of sand and gravel.[1] His line has continued to bear traffic; and it might have been hoped that the main lines of the development of modern science had been laid down upon firmer foundations and would require little but routine maintenance from the historian. On this view, the task of the historian of science is to assign dates to discoveries; and this once done properly would be done for ever. This conception of the history of science is in accordance with the idea that science is a collection of indubitable facts. But even values of physical constants, and the arrangement of species within genera, need to be revised from time to time; and in the more recondite and theoretical branches of science the revisions are more radical. Of the questions which the scientist puts to nature, some are difficult to answer and others difficult to ask; the historian must not ignore either the 'normal' or the 'revolutionary' questions, being careful not to adopt at the outset of his inquiry too narrow a definition of *science*. However widely he has read among secondary materials, he will then find in the primary sources the agreeable sting of surprise at the divergence between the unique occurrences there revealed and the tidy patterns which it is all too easy to impose upon them.

[1] J. A. Francis, *A History of the English Railway* (London 1851), I, 132; cf. N. W. Webster, *Joseph Locke* (London 1970), 41. For a view of science and its history as a collection of facts, see the address by Academician Kedrov to the XIII International Congress on the History of Science, Moscow, in press. On revolutions and normal science, see T. S. Kuhn, *The Structure of Scientific Revolutions*, 2nd ed. (Chicago 1970).

It is a mistake to suppose that there is one true history of science and that it is the task of the historian of science to write it down; rather, he should be looking for places from which a rewarding view may be obtained, in the knowledge that from any one viewpoint it will be impossible to survey the whole scene. Thus recent accounts of Humphry Davy[1] emphasise his chemistry as a reaction to that of Lavoisier, or show affinities between his conception of polar forces and that held by Romantics in Germany and in England, or reveal him as the apostle of applied science to the landed interest; these views, if well based upon study of the sources, must be not contradictory but complementary. There is much to be said for adopting the Coleridgean dictum that there is some truth to be found in any view, while admitting the futility of descending a well in order to see the world better. Thus even the still-popular conception of the opposition between science and religion, or the Church, can contribute to an understanding of the latter part of the nineteenth century when it was formed; although for other periods it would be little help in finding our way through our sources.

The study of our sources cannot but remind us that we are dealing with particular individuals living at a particular time, attempting to solve particular problems, and organising themselves into particular societies; it is not the task of the historian of science to deal with abstractions, and he should perhaps respond with coyness to the advances of the philosopher or sociologist of science. In the past, much history of science has been written by active or superannuated scientists who have often been anxious to legitimise a view of science, or even a particular scientific theory by showing that it has a long and honourable history. This can

[1] See the introduction by R. Siegfried to the reprint of Davy's *Works* (New York, in press); T. H. Levere, *Affinity and Matter* (Oxford 1971), ch. 2; M. Berman, 'The Early Years of the Royal Institution', *Science Studies, 2* (1972), 205–40. K. Coburn (ed.), *Inquiring Spirit* (London 1951), 412; J. W. Draper, *History of the Conflict between Religion and Science* (New York 1875), R. Hooykaas, *Religion and the Rise of Modern Science* (London 1972).

be characterised as whiggery, and it may go with an uncritical use of the sources; but we may describe such history more neutrally as 'applied' history of science, for it is surely desirable and laudable that scientists should wish to follow the development of concepts now important in their various disciplines. What is important is that there should be in addition 'pure' history of science, the concern of those wishing to discover what was seen as 'science' at different times in the past, what problems scientists faced, and how they tackled them.

The history of science has now emerged as an academic discipline, its followers forming a species of the genus 'historian' rather than 'scientist' or 'philosopher', and it is much to be hoped that more historians will turn their attention to this branch of their subject. So far it is the 'Scientific Revolution' of the sixteenth and seventeenth centuries which has received most attention from the critical historians, though economic historians have naturally looked at later technology. The science of the eighteenth and nineteenth centuries has been less studied, except for some dramatic episodes like the Darwinian controversy; and it has usually been presented in a manner daunting to those not trained in modern science. Scientists in these centuries were often their own popularisers, describing their discoveries for colleagues working in other sciences, and for the laymen who usually formed the majority in scientific societies and in the readership of many scientific journals. While therefore some aspects of the history of relatively-recent science can only profitably be studied by those who have studied modern science, there is plenty that is perfectly accessible to any historian, who will probably find the primary sources a good deal more attractive than most histories of science. There is something to be said for the history of science being pursued by those who have some experience of actual science—Grote might have written an even better history of Greece if he had visited the country—but the science of the past is not the same as the science of the present, and in studying it those with almost any background will have much to learn and something to add. The field is enormous, and rich harvests cannot but be gathered; provided only that historians pursue researches on which their sources can cast light.

The limits of our period in this study will be about 1660 and 1914. These include a great deal of science, with publications on the growth of geese from barnacles and on the antipathy of spiders to unicorn's horn at one end, and on quantum theory and Mendelian genetics at the other. But we are not here concerned with specific experiments, theories, or sciences, but with sources; and in this light the period does assume a certain unity. The most important point is the development about the 1660s of scientific societies which published journals making known the discoveries of their members. These journals were usually in the vernacular of the country in which they were published; in Germany Latin survived as the learned language through much of the eighteenth century, but in Italy, France and England scientific discoveries from the time of Galileo, Descartes and Boyle were published in the vernacular, and the journals accelerated this trend. Newton's extremely recondite *Principia Mathematica* of 1687 was an exception; but as a general rule only in medicine and botany did Latin continue as the normal medium of communication much beyond 1660. Authors like John Wilkins, first secretary of the Royal Society, and John Ray, the natural historian, even began to give English translations of Latin and Greek authors whom they quoted. The character of the sources therefore changes; knowledge of Latin is essential for the historian of the science of the Renaissance and of the early seventeenth century but not thereafter. In the first half of the seventeenth century we find chiefly translations of works of science into the vernaculars; in the second half such works are no longer being written in Latin.

The publication of journals by societies also brought into being a new source of science which steadily replaced less formal communications such as letters between men of science, or letters sent to a man of wide correspondence like Mersenne in order that their contents might be generally disseminated.[1] The historian of science of the period since the 1660s cannot but go

[1] A. C. Crombie is preparing an edition of Mersenne's correspondence.

through scientific journals, and from them he can often get an excellent view of the dialectical progress of a science. The scientific societies took various forms; they might be a group under the immediate protection of a nobleman, like the Academie del Cimento in Florence; a rather more open and egalitarian body under its own president, like the Royal Society of London; or a Department of State, like the Académie Royale des Sciences in Paris. Their journals also differed; the *Saggi* of the Academie del Cimento was not published until 1667, after the society had collapsed, and consisted of unsigned accounts of experiments done collectively over the few years of the society's life. The Royal Society's *Philosophical Transactions* contained for the most part signed articles setting forth some new observation, experiment, or theory; while the *Journal des Sçavans* concentrated upon reviews.

The *Saggi* have not been imitated, except perhaps in reports of committees usually on specific questions; but the two latter kinds of journal have been through our period—and indeed still are today—extremely important vehicles for the dissemination of science. In the second half of our period—that is, from the latter years of the eighteenth century—such general journals, covering the whole extent of natural knowledge, were joined by specialised publications concerned only with natural history, astronomy, or chemistry; and as specialised readerships grew, so the papers tended to become more technical. Journals published by societies were also joined at this time by proprietary publications of a general or specialised kind, published sometimes as a commercial enterprise but perhaps more often on behalf of a pressure group who were denied an outlet by the editors of existing journals.[1]

H. Oldenburg, *Correspondence*, ed. A. R. and M. B. Hall (Madison, Wisc. 1965–); the contemporary English translation of the *Saggi*, *Essayes of Natural Experiments*, tr. R. Waller (London 1684), has been reprinted (New York 1964, intr. A. R. Hall).

---

[1] S. Court, 'The *Annales de Chimie, 1789–1815*', *Ambix*, *19* (1972), 113–28; M. P. Crosland, 'Humphry Davy—an alleged case of suppressed publication', *British Journal for the History of Science*, *6* (1973), 304–10. On a proprietary journal, see A. Ferguson (ed.), *Natural Philosophy through the 18th Century*

Thus at the end of the eighteenth century one may find Lavoisier's disciples publishing in one journal, while adherents to the old theory of phlogiston published in another.

This brings us to our next point: that the scientific paper helped to bring into existence what has been caricatured as 'normal science', a somewhat humdrum process in which ordinary scientists, working along the lines laid down by those of greater originality, bring to light a steady succession of facts. Whereas in antiquity, in the medieval period, and in the Renaissance, scientific research seems to have been the pursuit of a few unusually able people, with spasmodic advances and long quiescent periods, the scientific societies and their journals both improved communications, and ensured that work would go on—not always perhaps very interestingly—on a wide front, as ordinary scientists, keeping up with their contemporaries, published their contributions in a steady succession of papers. Those advances which involve asking new questions, or looking at phenomena in a new way, demand more space than the editor of a journal can as a rule supply if they are to be argued convincingly; and in our period Newton, Lavoisier and Darwin made their impact through books.

While the scientific societies thus transformed the publication of science with their journals, they also helped to bring about the separation of science from other activities. In England and the U.S.A. by the end of the nineteenth century, and in other countries rather sooner, science had become a profession rather than an interest shared by various gentlemen, academics, clergymen, merchants and manufacturers. Academies of Sciences in most countries include among their members historians, archaeologists, philologists, philosophers, and others; but in 1901[1] the

(London 1972); but be warned that this reprints articles which first appeared in 1948, about the *Philosophical Magazine*.

---

[1] A. J. Meadows, *Science and Controversy* (London 1972), 228ff.; see I. B. Cohen and H. M. Jones (ed.), *Science Before Darwin* (London 1963). R. W. Church, *The Oxford Movement*, ed. G. Best (Chicago 1970), 20, 149; cf. T. F. Torrance, *Theological Science*, (London 1969). For cudgelling done 'scientifically', see H. Pottinger, *Travels in Beloochistan* (London 1816), 66.

Royal Society, to the fury of Lockyer, the editor of *Nature*, resolved to admit only those working in the restricted field to which we now apply the term 'science' and the British Academy was founded for those eminent in the 'arts', a word which earlier had meant techniques. By the end of our period, then, it had become clear what was science and what was not; the division into what has been called two cultures had in the English-speaking world come about, though whether this division is as interesting to explore as those between chemists, physicists and biologists is another matter. The word 'scientist' was invented only in the 1830s; before that date scientists were described as 'physiologists', 'naturalists', or 'natural philosophers', or often indeed simply as 'philosophers'. As this last title implies, they were expected to present and defend a world-view and often did so. The word 'science' began to acquire its modern meaning with the setting up of the British Association for the Advancement of Science in 1831; before that it had meant any organised body of knowledge. Thus in the *Encyclopedia Metropolitana* (1817–45), organised on a logical basis devised by Coleridge, the 'pure sciences' included grammar, rhetoric, law, theology and metaphysics as well as mathematics and logic; and the 'mixed sciences', those with an empirical component, included engraving, sculpture and fortification. Similarly, in his *History of the Oxford Movement*, Dean Church used 'science' to describe both the activities of the British Association and the dry and formal part of theology; in a younger man this latter use would by the 1880s have seemed old fashioned.

What one can then call the emergence of science as an activity distinct from other intellectual activities had happened by the end of the nineteenth century; and so had the emergence of science as a profession, with the introduction of courses in science at universities and technical institutions, the setting up of government laboratories, the passing of legislation on pure food and drugs, and the growth of science-based industries. But despite the numerical dominance of antiquarians, men of letters, and dilettanti, we can see in the early scientific societies the beginnings of what is loosely called the 'scientific community'. Papers for the journals were passed on to a referee to ensure their soundness, and to enable the editor to share responsibility; this should have meant that all experiments described had actually been performed and could be repeated, though a cursory examination shows that this was not the case.[1]

This community was moreover international; although science was a less cosmopolitan activity than is sometimes supposed, and national traditions in teaching and research were of great importance. From an early date, scientific societies had their foreign members or associates; and by the eighteenth century large-scale international cooperation on scientific expeditions had begun. The principle had become accepted that ships engaged in scientific work were not fair prizes in time of war; though this principle was sometimes honoured in the breach, as when Matthew Flinders was for years kept a prisoner of war in Mauritius on his return from surveying the coasts of Australia. When Davy was at the same period awarded a prize for his electrochemical discoveries by Napoleon, he went to France to collect it despite the war; there were grumbles about his lack of patriotism, but these were unjustified for Davy was as anti-French as anybody. In general, while scientists helped the war-effort by improving gunpowder and navigation, and by devising substitutes for products in short supply—oak-bark for tanning in England, for example, and sugar and coffee in France—no odium attached to those who were on the other side even when, like Lazare Carnot, Berthollet, and Laplace, they held high office in government. In 1914 things were different, and to be German was a sufficient reason for being expelled from the Royal Society.

The year 1914 thus marks a rupture in international relations between scientists which is reflected in the sources. There are

[1] See 'Sir' J. Hill, *A Review of the Works of the Royal Society* (London 1751); H. Woolf, *The Transits of Venus* (Princeton N.J. 1959); for Banks' efforts to get Flinders freed, see W. R. Dawson (ed.), *The Banks Letters* (London 1958), 328–32, and his supplementary paper, *Bulletin of the British Museum (Natural History), Historical Series, 3* (1962–69), 53–8. On 1914, see D. J. Kevles, 'Into Hostile Political Camps', *Isis, 62* (1971), 47–60.

other reasons besides which make this as good a time as any at which to break off. The professionalisation of science led also to its increasing fragmentation, with a rapid increase in the rate of publication, particularly in specialised journals; making it more difficult to see the wood for the trees, especially for those without a training in modern science. As the scientific community grew, so the small world of Victorian science changed into one where it cannot be taken for granted that scientists working in different fields will know or would understand one another; though this is something that one might illuminate by a close examination of sources. Again, it was about 1900 that modern science ceased to be essentially a European affair; Indians, Japanese and Chinese began making their contributions to theoretical science, as it seems their ancestors had to European science and technology, and the United States of America appeared belatedly as a major centre of science. There had been isolated American scientists of international reputation before, but most American scientists of the nineteenth century were natural historians engaged in describing the new country,[1] and America achieved technological predominance almost without significant research in pure science. American scientists became famous by about 1900 for the precision of their measurements, made with superb and expensive instruments; and the early years of the twentieth century mark the invasion of physics and chemistry by technology. Astronomy had always been expensive and had been lavishly supported by governments, as had geography, geology, and natural history; with the twentieth century, physics began to become expensive and to require the cooperation of a team as scientific exploration had in previous centuries. These changes in the character of science are associated with changes in the sources; and finally one might add that with the twentieth century we acquire the new source of living memory, and such novel materials as recordings of voices, and films.

[1] N. Reingold, *Science in 19th-century America* (London 1966), but see also his essay in G. H. Daniels (ed.) *Nineteenth-Century American Science: a re-appraisal*, (Evanston Ill. 1972), 38–62.

The history of science, like any other history, can only be written from the sources available; and if the sources do not enable us to answer some given question then there is no point in asking it. This does not mean that historians should not seek for new sources; there is great scope for ingenuity, and advances in our understanding of the history of science have been made, for example, by using the patterns tooled upon the leather of microscope tubes to date the instrument, as well as by the close study of letters, minutes or account books previously unknown or ignored. Indeed it is a stimulating feature of the history of science that so much is still in doubt, and that there is a mass even of published material that remains unused and even unread. But however full our sources might be, there would remain questions which are unanswerable, or to which unambiguous answers cannot be given. One of these questions is the one with which historians of science have sometimes chiefly preoccupied themselves; determining who discovered what. Even in matters of geography, it is often difficult to decide who found what: as Boorstin has told us,[1] settlement in America usually preceded survey; and whether Columbus can really be said to have discovered America since he supposed himself to have arrived at the dominions of the Great Khan is a question which different people could answer in different ways.

In the sciences we find both these difficulties. The magnetic compass does not point to true north, and its variation from true north is different in different places and changes with time. This was known to makers of compasses, who allowed for it in their instruments, long before it was discovered by the scientist Henry Gellibrand of Gresham College. In the same way, engineers knew that high-pressure steam engines were more economical than those working through a smaller range of temperature long before the Second Law of Thermodynamics was propounded. More complex questions are raised when we look at more recondite discoveries; to describe a country as 'America' might be taken as implying that it is a new continent, and in the sciences almost

[1] D. Boorstin, *The Americans*, II (New York 1965), ch. 28.

all terms are theory-laden in this way. We need not consider whether all facts are theory-laden—that is a matter for the philosopher—but this aspect of terminology reflects the way in which sciences are organised bodies of knowledge in which observations are interpreted. Thus[1] the Swedish chemist Scheele discovered that when 'Marine acid' is 'dephlogisticated' by levigated manganese, a yellowish-green choking 'air' is emitted; he therefore called this air 'dephlogisticated marine acid'. Berthollet, a disciple of Lavoisier who rejected phlogiston, reinterpreted the reaction; for him, 'muriatic acid' acquired oxygen from the manganese peroxide, and what was given off was 'oxymuriatic acid gas'. Humphry Davy could find no oxygen in the gas, rejected Lavoisier's doctrine of acidity, and decided that what has since been called 'hydrochloric acid' had its hydrogen removed to yield the element 'chlorine'. The question 'who discovered chlorine' has no simple answer; Scheele prepared the substance to which we now give the name, Berthollet discovered its bleaching properties, and Davy named it and proposed that it was an element. This question was no clearer to contemporaries than it is to us; Davy indeed generously credited Scheele with the discovery, discerning his own views beneath Scheele's terminology.

Contemporaries took no notice of this modesty, or appeal to authority, as they had not of Newton's view that in his theory of gravity he had recovered the esoteric Pythagorean doctrine concealed beneath their talk of the Harmony of the Spheres; and in this they showed their historical sense. To ask questions like 'who discovered chlorine, or oxygen' is futile, and cannot but obscure from our view what the scientist was trying to do and why his contemporaries were, or were not, interested. Indeed there is much to be said for the historian of science divesting himself of the idea of progress, as other historians have done; it may be that science has made progress, but nothing is gained by assuming it. Instead of asking how somebody fitted into the progress of science, one should ask what he observed and how satisfactorily

[1] This episode may be followed through the papers reprinted and translated in *Alembic Club Reprints*, *9* and *13* (Edinburgh 1894, 1897).

he explained it. Thus traditionally Wöhler was credited with overthrowing vitalism by his synthesis of urea in 1827;[1] whatever 'vitalism' may have been it was far too complex to be overthrown by one experiment, and if instead of being mystified as to how vitalists continued even in high places for many years longer, we see Wöhler's synthesis as casting light on the arrangements of atoms in compounds then we gain useful insights into the chemistry of his day, which acquires a new coherence for us. In other words, we get a good view instead of a restricted one; and we can make sense of more of the sources.

Scientists have tended to look upon their discoveries as their property; and therefore to get involved in priority-disputes. Since membership of prestigious societies, patronage and financial security have often depended upon the making of original discoveries, we need not be surprised at this, although often the best established scientists have been vehement defenders of their own priority in some discovery of minor importance. That simultaneous discoveries were made in the sciences was known even to the Crown Prince of Burma in the 1850s.[2] The historian with access to the documents on both sides may be better placed than were contemporaries in deciding whether the simultaneous discoveries were genuinely independent, and which scientist was just ahead in discovery or publication. But the historian's concern is different; he is recording, and as far as possible accounting for, an historical phenomenon rather than deciding who deserves a medal. He will therefore look at the sources to see what the different people involved were looking for, how they described their discovery, and how far they surprised their peers and

[1] J. H. Brooke, 'Wohler's Urea and its vital force?', *Ambix*, 15 (1968), 84–114; and his 'Organic Synthesis', *British Journal for the History of Science*, 5 (1971), 363–92.

[2] H. Yule, *A Mission to the Court of Ava in 1855*, ed. H. Tinker (Kuala Lumpur 1968), 125. On Mendeleev and the others, see J. W. van Spronsen, *The Periodic System of Chemical Elements* (Amsterdam 1969); the papers are reprinted in facsimile in my *Classical Scientific Papers: Chemistry, 2nd series* (London 1970). A study of contemporary discovery is M. Grosser, *The Discovery of Neptune* (Cambridge, Mass. 1962).

convinced them that it was interesting. While recording them, then, the historian can keep out of priority disputes; his task is to see who was given credit by contemporaries, and why—recognising that credit does not always go to the most deserving. Thus in proposing the Periodic Table classification of chemical elements, Mendeleev was not unique; at the same time and even before, chemists in England, France and Germany had put forward very similar schemes. But Mendeleev obtained the credit because he saw further than the others what use might be made of the Table, and staked his reputation upon it. This credit is shown by the wide use and translation of his textbook, by the references in other textbooks, and by the memberships, medals, and invitations to lecture given him by scientific societies.

Mendeleev was not honoured by the Academy of Sciences in his own country, and this can serve to remind us how often the sources reveal that what appear to be questions of intellectual history resolve themselves into questions of academic politics, or other kinds of politics. Thus the view that gases are composed of particles in rapid motion—the kinetic or dynamical theory of gases—was proposed at various times in the nineteenth century but was not taken seriously until Clausius and Maxwell, both of them established and important physicists, forced it upon the attention of their contemporaries.[1] Voltaire's propagation of Newtonian physics in France was a part of his anglophilia; and similarly the reception of Lavoisier's chemistry was connected with a liking for or abhorrence of things French. Although insights can be gained this way, there seem to be very few cases where the reduction to academic politics leaves no residue. Thus Charles Babbage's *Decline of Science in England* (1830) must be taken as propaganda in favour of more power and honour for pure and applied mathematicians, but some of the accusations there made against the Council of the Royal Society are telling and

[1] R. G. Brush, *Kinetic Theory*, 2 vols. (Oxford 1965–66; reprints some documents); others are in my *Classical Scientific Papers; Chemistry* (London 1968). Darwin's scientific morality comes under attack in C. D. Darlington, *Darwin's Place in History* (Oxford 1969).

raise general questions on the practice and administration of science. People accept or reject theories or views of science for a variety of reasons, which the historian should as far as possible enumerate in the cases he is studying; he does not need to make moral judgements, though he should take those of contemporaries into account.

As intellectual questions may in varying degree resolve themselves into political ones, so questions of truth may often usefully be rephrased in terms of function. This is especially true with what some people choose to call pseudo-science. The mid-seventeenth century in England was a period when great numbers of alchemical and astrological works were published; in the late eighteenth century came mesmerism, then phrenology; while in the latter part of the nineteenth century there was a great eruption of interest in psychical phenomena.[1] Those who were counted among the eminent scientists of the time took a serious interest in these things; one must not ask the question 'why did people believe such nonsense?', for to this question the sources can give no answer and all that results is the warm glow of feeling that we are wiser than our fathers. If however we follow Keith Thomas and ask 'why did people consult astrologers?', we may find as he has done, that the sources enable us to answer that question; that an astrologer provided what might now be called counselling. Thomas also shows how within its own framework astrology was irrefutable; a failure in prediction merely indicated the need for more research, and it was only with the emergence of a mechanical world-view that astrology became unreasonable. One can lay it at the door of whig historians of science that by ignoring the nonsense they have adopted too narrow a view of their subject; it might be better to take as one's definition of science 'what Fellows of the Royal Society and their equivalents do' at

[1] See A. Debus' introduction to the reprint of E. Ashmole, *Theatrum Chemicum Britannicum* (1652) (New York 1967); A. Gauld, *Founders of Psychical Research* (London 1968); F.-A. Mesmer, *Le Magnétisme Animal*, ed. R. Amadou (Paris 1971); K. Thomas, *Religion and the Decline of Magic* (London 1971); F. Yates, *Giordano Bruno* (London 1964).

any given period. For the Renaissance and early seventeenth century, this balance has in recent years been amply redressed; for the eighteenth and nineteenth centuries some corrections to the received view could still be made.

If the frontiers between science and activities now less reputable are hard to draw, the sources giving us little support, those between different branches of science often cannot be delineated at all without anachronism. It is perhaps an indication of the immature state of the history of science that at international Congresses the various 'sections' discuss the history of a particular discipline rather than the science of a particular period. This is a sensible way to proceed if one is doing 'applied' history of science, that is, following the evolution of concepts now important; but it will not do for the serious historian of science to study only the history of one science, at any rate in the period we are considering. Study of the sources shows the impossibility of doing this, for many scientists worked across the whole spectrum of sciences, and fashionable concepts from one science were rapidly adopted in others—sometimes in place of concepts that seem to us more suitable. Because scientists and historians of science have looked back to find men they can categorise in modern terms, those who worked over what now seems a wide field are less remembered than contemporaries of narrower range; and both are often remembered for things which did not strike them or their contemporaries as their most important work. Thus (in contrast to Dalton) Hooke, W. H. Wollaston, and John Herschel have been less studied by historians of science than their importance in their own day warrants; and when in various histories of different sciences we do find mention of their works we cannot see the man whole, and it looks as though a number of discoveries were made by different people of the same name living at the same time. To recover contemporary judgements, to see science as a continuing activity, and to make sense of the sources, one must study a wide spectrum of science in a relatively brief period. Historians have tended to drill a small hole down from the present through the strata of history; they would be well advised instead to look much more closely at the contents of one particular stratum.

The distinction between internal and external history of science, which is sometimes insisted upon, similarly begins to appear artificial when we look at the sources. It has been suggested that the historian of science looks at the development of science while other sorts of historian look at its context; but to examine documents or apparatus in ignorance of their background, or the context of science in ignorance of the science itself, is not an exercise one would want to recommend. *De facto*, there often is this division; but the sources do not warrant it. Some sources are considerably more technical than others, and sources are technical in different ways; to follow Newton's researches into the chronology of the ancient kingdoms, for example, demands a different expertise from that required to read his mathematical papers.[1] Different historians are equipped to take advantage of different sources and will weight them differently; but any who were to suppose that Newton's career was to be completely described in terms of the economic situation of his day, or contrariwise that Newton read nothing but mathematics, would be flying in the face of the sources. There is surely a continuum, from the formal paper in learned journals to the dinner party invitations, account books, and artefacts which may also be sources for the history of science; there is no easy way of dividing the sources into internal and external, and the decision as to what sources are of little moment is one that can only be made after a survey in each particular case. What must be remembered is that every source has its context; every document or object was made with at least one end in view and cannot be comprehended in isolation; and while the historian cannot be omniscient he should try to make his knowledge both wide and deep.

Other frontiers perhaps more convenient than natural are those

[1] See F. E. Manuel, *Isaac Newton, Historian* (Cambridge 1963); D. T. Whiteside (ed.), *Mathematical Papers of Sir Isaac Newton* (Cambridge 1967–). In general, see W. Cannon, 'History in Depth: the Early Victorian Period', *History of Science*, *3* (1964), 20–38.

which separate the history of science from the histories of medicine, mathematics, technology and education; the historian should only respect these frontiers when it suits him to do so. Medicine was for long the only university training involving experimental science, and for longer the obvious profession by which a biologist or chemist could support himself; the connexion between mathematics and science has always been close; and much educational controversy has been concerned with attempts to introduce science into curricula.[1] Relations between science and technology have also been sometimes very close; recent work has brought out some fruitful periods of close cooperation between what we would now call scientists and technologists, and has indicated that many people cannot be classified as one or the other.[2] But it is worth remembering for instance that the most important figures in the establishment of the Principle of Conservation of Energy—Mayer, Joule and Helmholtz—were not in any sense engineers; and that recent close studies of the Manchester Literary and Philosophical Society seem to indicate that those manufacturers who belonged to it did so in the hope of improving their conversation and status rather than their machinery, being most attracted to the sublime science of astronomy. These marches between science and technology can be profitably explored by the historian, as may such more familiar borderline territories as the science-based industries, and the manufacture and use of scientific instruments and apparatus.

This last study leads us not only into the world of precision

[1] See M. Kline, *Mathematical Thought from Ancient to Modern Times* (London 1972); L. S. King, *The Road to Medical Enlightenment* (London 1970); A. Debus, *Science and Education in the Seventeenth Century* (London 1970); W. H. Brock is writing a book on later science teaching.

[2] R. E. Schofield, *The Lunar Society of Birmingham* (Oxford 1963); A. E. Musson and E. Robinson, *Science and Technology in the Industrial Revolution* (Manchester 1969); E. Robinson and D. McKie, *Partners in Science* (London 1970); D. S. L. Cardwell, *Technology, Science and History* (London 1972); A. Thackray is working on the prosopography of the Manchester Literary and Philosophical Society in Dalton's day; D. Landes, *Unbound Prometheus* (Cambridge 1969).

instrument making, enabling us to assess the skills, incomes and perhaps social position of men who often themselves made observations or wrote textbooks of some importance; but also casts light on the accuracy of the observations available at a given time. Kepler was fortunate in having available observations of planets too accurate to be accounted for in terms of circular orbits, but not so accurate as later observations which would have shown divergencies from the elliptical paths which he postulated. Again, science is sometimes supposed to consist of series of verifiable or falsifiable propositions; but an hypothesis so apparently falsifiable as that of Prout, who suggested that the atomic weights of all elements might be integer multiples of that of hydrogen, kept the best analysts in Europe in controversy for nearly twenty years, and in modified form survived even longer. When observations are made at the limits of accuracy of the apparatus, and when there is little theory available to guide the experimenter, it is hard to produce results that will convince all contemporaries. Often cases which seem to us as though they must have been cut and dried, turn out on examination of the sources to have been difficult and puzzling to contemporaries. And we find too how much time had to be spent on such relatively banausic activities as rating chronometers to make what seems a simple determination of longitude, or in setting up the apparatus to perform some analysis.[1]

The examination of instruments when they survive, of tables of observations made with them, and of the methods used to produce an 'average' value from a number of readings of the

[1] On Kepler, see C. D. Hellman in A. Beer (ed.), *Vistas in Astronomy*, 9 (Oxford 1968), 43–52; on Prout's Hypothesis, W. H. Brock's papers, *Annals of Science*, 25 (1969), 49–80, 127–37; the primary papers are reprinted in *Alembic Club Reprints*, 20 (Edinburgh 1932), and in facsimile in my *Classical Scientific Papers; Chemistry, 2nd series* (London 1970). On chronometers, see H. Quill, *John Harrison* (London 1966); M. Flinders, *A Voyage to Terra Australis* (London 1814), I. 255ff., and W. E. Parry, *A Voyage for the Discovery of a North-West Passage* (London 1821), Appendix I, give details of rating them; see also *The Journals of Captain James Cook*, ed. J. C. Beaglehole, 3 vols. in 4 (Cambridge 1955–67). S. Bradbury and G. L. E. Turner (ed.), *Historical Aspects of Microscopy* (Cambridge 1967).

instrument, can give us an idea of the accuracy available at a given time. It may also help to indicate whether the science was detained at what has been called a 'technical frontier'. Thus it appears that embryology and bacteriology had to wait upon the improvement of microscopes; early microscopes both distorted the image because of the form of their lenses and produced coloured fringes all around it because these lenses were made of only one kind of glass. As in the nineteenth century these defects were overcome, so rapid progress took place in sciences for which a good microscope was essential. In a similar manner the conquest of scurvy, and the simultaneous improvement of instruments and tables for lunar observation and development of the chronometer, made it possible to settle the question of whether there was a great southern continent or not. Conversely, Copernicus advanced astronomy not by new observations but by handling his elderly data in a new way. Pure examples of technical or intellectual frontiers are rare; for the microscopic observations needed to be interpreted, Cook and his companions were not merely answering old questions, and Copernicus made a few observations and stimulated others to make many more with improved instruments. It may be better to stick to the less-formal distinction between questions difficult to answer and difficult to ask. Often the technological frontier is crossed in a way that those who first perceived it could not have imagined. Thus chemists of the nineteenth century tried to relate atomic arrangements to crystal structures, and early in the present century it proved possible to do this by interpreting photographs taken with the newly discovered X-rays; and Babbage in the 1820s designed clockwork calculating engines which with the rise of electronics 120 years later proved the basis of our computers.[1] Such reflections may be interesting, but they are hardly history; the historian must be concerned with how people used and improved their apparatus, and with whom they cooperated in this, in order to test their hypotheses. To do otherwise is to give

[1] C. J. Schneer, *Mind and Matter* (New York 1969); P. and E. Morrison, *Charles Babbage and his Calculating Engines* (New York 1961).

a specious modernity to remarks taken out of context.

Specious modernity is particularly hard to avoid when translating works; the translator must have a command in both languages of the scientific terms and concepts of the day, and because this is unusual contemporary translations are as a rule to be preferred even if they may appear—or did appear—inaccurate or clumsy. We thus avoid anachronism; and moreover we find ourselves using what contemporaries used. Translators and editors were often incompetent or cavalier, but what appeared in print was what the writer was taken to have written. This brings us to the vexed question of influence. Partly because historians of science have looked so much at the development of concepts and at new discoveries, they have acquired an almost-literary obsession with originality; and have therefore sought to find who influenced whom to see how original a discovery was and to whom it must ultimately be referred. This can be straightforward; we may find a reference in a paper or book, or a statement in a diary or letter, in which our author tells us that he has just met somebody, or read something, and got some new idea or information thereby. In the absence of such direct evidence—and even in diaries and letters people do not always tell the whole truth—influence is more difficult to prove. We may find that our author owned or borrowed a book containing the doctrines that we think influenced him; but we do not know unless he annotated it that he read the book at all, let alone the passage in question, or that he found in it the doctrine which so impresses us—he may have been looking for something quite different. Thus Mendel's paper on heredity remained almost unknown for more than thirty years;[1] and those who did refer to it in bibliographies seem to have supposed that it was a horticultural work, on the cultivation of peas. The English translation of Mendel's paper did indeed appear in 1901 in the *Journal of the Royal Horticultural Society*; but by then its wide importance was evident.

[1] R. Olby, *The Origins of Mendelism* (London 1966), discusses the reception of the theory. Translators might find useful the bilingual works on natural history which were a feature of eighteenth-century publishing.

It is, in questions of influence, very difficult to prove a negative: thus we know that Newton read Descartes because we have his notebooks; but we do not know that he read Galileo, and an ingenious case has been made out from citations made, and not made, by Newton for thinking that he did not.[1] On the other hand, nobody with Newton's interests could have escaped the influence of Galileo. More doubtful, for example, are claims that Davy and Faraday were influenced by Schelling and his disciples in working towards a world view in which polar forces were predominant. Neither Davy nor Faraday spoke German, and such remarks as they made about *Naturphilosophie* were uncomplimentary; but we are sometimes rude about those from whom we have borrowed, especially if we have borrowed heavily. Coleridge has been invoked as the connecting link; he and Davy were at one time very close friends, and he might well have passed on the latest German theories of matter and force. But Coleridge was not one to pass on a doctrine exactly as he received it: and the work of Davy and Faraday had an experimental basis and led to predictions that were confirmed; so that their views were in the end generally accepted in the scientific community as those of Schelling and Ritter had not been. There may have been direct or indirect influence; but the sources do no more than indicate parallels or possibilities and leave us in the frustrating position of being unable to prove anything either way. Since Davy and Faraday clearly added much to whatever they might have received from Germany, it is not evident that proving or disproving influence would anyway be very interesting, except insofar as it illuminated relations between England and Germany in the Romantic Period.

Rather than seek for these influences, the historian of science would be well advised to look for the norm at any given period; to

[1] A. Koyré, *Newtonian Studies* (London 1965), essays III and IV; L. P. Williams, *Michael Faraday* (London 1965), 137ff; T. H. Levere, *Affinity and Matter* (Oxford 1971), chs. II and IV; but see B. S. Gower, 'Speculation in Physics: the History and Practice of *Naturphilosophie*', *Studies in the History and Philosophy of Science*, *3* (1973), 301–56.

see what was taken for granted, how people were trained, and what relationships prevailed between scientists. On these points sources can be helpful, though we may find ourselves devoting less time to famous works by great men and more to textbooks, syllabuses and the records of societies. After all it is only when we have an idea of the norm that we can assess the great men; and until we have such an idea we are all too prone to over rate rather than to understand the scientist whose work we are examining. But more than that, the recovery of the norm is itself interesting and must be the primary task of the historian. The science of the 1670s, 1770s and 1870s was different, and differed from the science of today; and the differences do not simply consist of a few or many discoveries added in each century by great men. There have always been a few who have sailed in strange seas of thought, alone; but to most people science at any period has consisted of ordered collections of facts, and of generalisations and assumptions usually taken for granted rather than stated. To find the assumptions set out, one must often turn to reviews, essays or lectures; to works of natural theology or philosophy, and even to didactic poetry. It is as attempts to establish the norms in different periods and different countries that the more interesting recent works in the history of science can best be seen;[1] and in such attempts we may be led to a wider and fuller use of the sources, and to the use of fresh sources.

Thus it has been suggested that some scientists made their discoveries in part because being isolated they were ignorant of the norm, or the accepted way of attacking problems. Benjamin Franklin, for example, is said to have made his electrical discoveries through a 'naïve insight'; cut off in Philadelphia from most scientific literature, he was able to approach the problem afresh.[2] Dalton has been the subject of a more searching investigation, and it does look as though he was able to formulate his

[1] See for example, G. W. F. Hegel, *Philosophy of Nature*, tr. and ed. M. J. Petry, 3 vols. (London 1971); *The Papers of Joseph Henry*, ed. N. Reingold (Washington D.C. 1972–).

[2] D. J. Boorstin, *The Americans* (New York 1958), I, ch. 39; for a fuller

crude but useful form of atomism because he was in Manchester unaware of the high-flown and subtle arguments about the nature of matter into which his contemporaries in London and Paris could not but be drawn. Largely for this reason, his contemporaries accepted his laws of chemical change but rejected the theory which had guided him to them; and Dalton's atomism only became generally acceptable about half a century after its author's death. Dalton also entered chemistry as a meteorologist; and indeed many advances have followed upon an able man trained in one field entering another. Thus Faraday entered electricity with the training of a chemist rather than of an applied mathematician; while conversely Maupertuis and R. A. Fisher were mathematicians who made fruitful incursions into the biological sciences.

At any one time, therefore, there is not simply one norm or paradigm. Not only do we find that those working in different fields have been taught to think differently, but also that there are divisions between those working in the same field but trained in different countries, or even only in different universities. Thus in the history of the concept of energy, various traditions seem to have been characteristic of various countries throughout the eighteenth and early nineteenth century, and advances seem to have been brought about by those who succeeded in making themselves familiar with more than one of them.[1] A similar story can be told about theories of chemical combination and of the structure of chemical compounds in the nineteenth century, where visits to another country by able men often led to a synthesis of two traditions and a wider understanding of the phenomena. The interchange was not always successful; thus part of

account of Franklin's work, see I. B. Cohen, *Franklin and Newton* (Philadelphia 1956). On Dalton's atomism, see A. Thackray, *Atoms and Powers* (Cambridge, Mass. 1970), ch. VIII. P.-L. M. de Maupertuis, *The Earthly Venus*, tr. S. B. Boas, ed. G. Boas (New York 1966); R. A. Fisher, *Natural Selection* (Edinburgh 1929).

[1] W. L. Scott, *The Conflict between Atomism and Conservation Theory* (London 1970); C. A. Russell, *The History of Valency* (Leicester 1971); A. J. Meadows, *Science and Controversy* (London 1972), 77, 292ff.

the reason for Norman Lockyer's difficulty in getting his, admittedly bold, astrophysical ideas accepted seems to have been that most astronomers in England were Cambridge men, trained as applied mathematicians, and he was not. And wherever there was communication across a linguistic barrier, there was an increased possibility of misunderstanding; sometimes this was advantageous,[1] and the concept of 'creative misunderstanding' has been evolved to describe how for example Kant drew from a review of a curious work by Thomas Wright his nebular hypothesis of the origin of the solar system.

These various traditions can be identified from the sources, when we find series of papers being produced which contain common pre-occupations and employ common terminology; evidence from these formal productions will probably be backed by correspondence, or evidence that those concerned knew each other fairly well. Terminology can be particularly important, because if two people are accustomed to different jargons, and the different way of looking at things thus implied, they will only understand each other imperfectly. Again, a terminology which appears modern can deceive the historian by giving a specious modernity to the sources; this can only be avoided by extensive reading and the use of dictionaries of the period. We are perhaps apt to suppose that scientific terms are more rigorously defined and used than are ordinary words. This may be true of severely technical terms like 'isotope' or of units like 'volt', but such more general and interesting words as 'evolution' and 'energy' have undergone considerable semantic change and were widely used at times when their meaning was fluid. Thus 'evolution' is not used in the first edition of the *Origin of Species*, the term 'development' being preferred probably because 'evolution', used of the growth of an embryo or the unfolding of a bud, carried in 1859 more overtones of predictable progress than Darwin wished. Again 'energy', and the associated terms 'force' and

[1] T. Wright, *Second Thoughts*, ed. M. A. Hoskin (London 1968), 9; and the criticism by G. J. Whitrow in *Kant's Cosmogeny*, tr. W. Hastie (New York 1970), xxviii.

'power', gradually lost their eighteenth-century connotations with people or immaterial agencies and did not refer clearly to definite mechanical quantities until the middle of the nineteenth century. The man who is 'ahead of his times' may be using such words in a more modern sense than his contemporaries did; and if we have definite evidence from his letters or notebooks then that is interesting to know. But his audience or readers would have taken his words in their accepted sense; and what a man was taken to mean is just as important to the historian as what in fact he may have meant, if not more so. Contemporary translations can be a help in deciding what were accepted senses of words; for we sometimes find that a competent translator has not rendered a word by what seems to us the obvious equivalent. But one should inquire into the circumstances of the translation before giving too much weight to such evidence.[1]

As part of our effort to discover what were the norms, we shall find that different sciences have been predominant at different periods. Thus in the early years of the twentieth century historians to whom 'science' meant physics were critical of the Royal Society under the forty-year reign of Sir Joseph Banks because they believed that this was an amateur regime. But close study of Banks' correspondence and journals shows that he was a very competent natural historian indeed, and a powerful patron of natural history and of geographical and agricultural science. In the twenty years either side of 1800, and indeed beyond, 'science' did not mean physics, and to have a natural historian in command of the Royal Society was not surprising.[2] In the U.S.A. this state of affairs lasted longer, because Americans had a continent to explore. In deciding what science was predominant,

[1] See e.g. B. Linder and W. A. Smeaton, 'Schwediauer, Bentham, and Beddoes: translators of Bergman and Scheele', *Annals of Science*, *24* (1968), 259–73.

[2] A. M. Lysaght, *Joseph Banks in Newfoundland and Labrador, 1766* (London 1971), N. Reingold, *Science in 19th-century America* (London 1966); G. S. Ritchie, *The Admiralty Chart* (London 1967); W. Cannon, 'History in Depth', *History of Science*, *3* (1964), 31, 35.

we have various criteria, one of which is to see who held the powerful positions in the scientific world, received the most honours, and perhaps commanded the highest salaries or drew the largest audiences. In Britain in the century following the entry into the Pacific of Byron, Wallis, Cartaret and Cook, expeditions of an increasingly scientific nature received considerable government support; the most dramatic of these being the voyages to the polar regions under Parry, Franklin, Beechey and Ross. Voyages of discovery carried naturalists and astronomers, and naval officers and surgeons turned themselves into excellent scientists. When we include these activities, as we should, under science we find apparently that the Duke of Wellington's government spent a higher proportion of its revenue on science than later nineteenth-century British administrations. The discoveries of the natural historians enriched the cabinets of the curious, who included numerous men of power and wealth.

Powerful support is another criterion of predominance, therefore; and this is something which can be discovered from the sources. More elusive is intellectual interest: there was immense popular enthusiasm over the return of expeditions, but great crowds flocked also to lectures on medicine and on chemistry, and when Davy visited Dublin there was a black-market in tickets for his lectures on electrochemistry. Chemistry and astronomy did probably pose the most exacting intellectual problems in early nineteenth-century science, and had connexions with determinism and materialism which made them interesting sciences to the philosopher. This kind of predominance has to be assessed from contemporary publications; and this too is something which can be done. The period of Banks' ascendancy in the Royal Society, and of the ferment in astronomy and chemistry, is also one in which the sciences of animal magnetism, physiognomy and phrenology were much before the public mind; the limits of science have varied, and the historian must not assume that doctrines which to him seem ridiculous could not have been held by intelligent people in the past. Phrenology, for example, was taken very seriously, as a deterministic psychology, among free-thinkers and educationalists in the first half of the nineteenth century.[1]

The 'decline of science in England' which Babbage deplored in 1830 meant essentially a change in those who held power from Newton the astronomer to Banks the natural historian. But different countries have risen and declined in scientific power as in other kinds of power; France, for example, was under the First Empire a great power in the world of science, a centre for advanced research in numerous fields, but by the time of the Second Empire she counted for much less.[2] Historians of science have given reasons more or less plausible for this, uncertain whether the decline was absolute or relative, or indicated specialisation by the French in certain fields to the neglect of others which attracted attention in England and Germany. One finds not infrequently the prophet not without honour save in his own country; the work of Willard Gibbs was not taken seriously in the U.S.A. until Maxwell and Ostwald made its importance clear, and Arrhenius only achieved recognition at home after receiving it abroad. In both cases this was because their researches were abnormal to their compatriots, straddling an academic frontier which in their own country seemed natural and fixed; while in other countries, with other norms, their work was a welcome contribution. An able young man looking for a reputation may choose a field cultivated abroad and neglected at home. Rather than being too concerned about upsurges or declines of science in different countries, the historian might find that his sources gave clearer answers to the question 'what sciences were being actively pursued in different countries at different times?',

[1] G. S. Haight, *George Eliot* (Oxford 1968), 37; G. Combe, *The Constitution of Man*, 8th ed. (Edinburgh 1847), reprinted Farnborough, 1970.

[2] See the circulated papers for the Conference of the Past and Present Society and the British Society for the History of Science, on *The Patronage of Science in the 19th Century*, in London, 14th April 1972; the paper by R. Fox on science in France. On Gibbs, see N. Reingold, *Science in 19th-century America* (London 1966), 315ff.; H.-G. Körber (ed.), *Aus dem wissenschaftlichen Briefwechsel Wilhelm Ostwalds* (Berlin 1961). Cf. the situation in Italy at the end of our period: E. Segrè, *Enrico Fermi* (Chicago 1970), 40.

and gave some indications of the reasons for the differences.

Such studies give little support for the view that the history of science might form a part of a 'science of science' in which predictions in sufficient detail to be useful might be made from what happened in the past.[1] Attempts to quantify the development of science indicate an exponential growth leading to the lunatic conclusion that at some time there will be more scientists than people, or to the sensible but unhelpful view that things cannot go on thus indefinitely. But these quantitative exercises often beg all the interesting questions raised by the sources. Unless it is clear who should be included among scientists, and whether before the nineteenth century this is a classification that makes sense, it is vain to try to enumerate them. The gnat who asked Alice whether the insects in her country answered to their names, and if not what was the use of their having names, can teach us a lesson here; there is little point in describing as scientists those who would not have thought of themselves as being in such a category. This problem is made far more acute when one tries to determine quantitatively how many scientists were puritans, for example, for one then has two categories to apply; and the widely-held view that dubious data fed into a computer will yield accurate and important results should be combated. At least in the present state of the history of science it is to the sources rather than to statistical manuals that the historians should turn; and we must now embark upon our general survey of these sources, beginning with histories of science.

[1] M. Goldsmith and A. Mackay, *The Science of Science* (London 1964); D. J. de S. Price, *Little Science, Big Science* (New York 1963); R. K. Merton, *Science, Technology and Society in 17th-century England* (New York 1970); C. N. Hinshelwood in D. S. L. Cardwell (ed.), *John Dalton and the Progress of Science* (Manchester 1968), cautions us about quantification. See too R. Floud, *Quantitative Methods for Historians* (London 1973). L. Mulligan, 'Civil war politics, religion and the Royal Society', *Past and Present*, 59 (1973), 92–116. On the hazards of computer models, see H. S. D. Cole et. al., *Thinking about the Future* (London 1973).

CHAPTER 2

Histories of Science

Since it is to the secondary sources that those intending to enter the history of science will probably first turn, it is appropriate to begin our survey with them. Histories of science fall into various groups: there are the formal histories, covering extended or limited topics and periods; the historical introductions appended to many works of science, or to courses of lectures on science; and the memoirs, obituaries, and biographies of men of science. Histories of science in these various forms have been written throughout our period and on to the present day; what they all have in common is that they were written with some end in view. Where the objective is very clear, there are few difficulties; but in all histories the preconceptions of the author and of the times are embodied, and the definitive account of events, written without bias, is a chimaera. This is not to say that historians have been and are dishonest; we must look for dishonesty and incompetence and we may find it, but differences in judgement and interpretation are unavoidable and give interest to history. Faced with any source, primary or secondary, we must ask by whom, when and why it was produced.

There are indeed some histories of science which aim at chronicling everything; but these cannot but be excessively dull—after all, history was the province of a muse and a work of history should be a work of art—and at best only provide data for the historian to work upon.[1] Even these data have been passed through someone who believes that the historian's task is to record events in chronological order; and who, since he is not omniscient

[1] Examples of this genre are L. Thorndike, *A History of Magic and Experimental Science*, 8 vols. (New York 1929–58); and J. R. Partington, *A History of Chemistry*, 4 vols. (London 1961–70); J. Needham, *Science and Civilisation in China* (Cambridge 1956-) is rather different.

and has not infinite space at his disposal, clearly cannot even record everything. More seriously, in the amount of space given to various scientists, sciences, or divisions of science, and in the authorities consulted, the author has been guided by his own views of what science is and was. From such a compilation we get a muddier view of the science of the past than we do from a lively book written from a recognisable viewpoint. The historian should not turn first to such multi-volume histories, however accurate; but he may later use them with care to find dates of birth, death, or election to societies—though if anything hangs upon these dates they must be confirmed from primary, or as far as possible contemporary, sources. The production of such annals is one of those activities that makes one at first wonder how it is done, and then wonder why; though there would always be room for extended works, of a critical and analytical kind, the history of science cannot in the present century boast of a Clarendon, a Gibbon or a Macaulay. The one exception to this possibly melancholy picture is Joseph Needham, whose massive survey of the exotic science and civilisation of China, which he has shown to be less exotic than had been generally supposed, breaks new ground which we may hope will soon be further occupied; for in the history of science it is good that any territory should be worked by a number of people, having rather different interests.

Because one person cannot be omniscient, those who hope for a definitive history sometimes pin their faith upon a collective work in which each chapter is written by an expert. At best, as in our field with the *Oxford History of Technology*,[1] this can produce an interesting series of brief essays very valuable for anybody wishing to study the history of science; but even here the essays are uneven and show a lack of overall direction, so that the work could never be read right through and the effect is often antiquarian rather than historical. Where the authors are

[1] C. Singer, E. J. Holmyard, A. R. Hall, T. I. Williams (ed.), *A History of Technology*, 5 vols. (Oxford 1954–58); an example of what to avoid is R. Taton (ed.), *General History of the Sciences*, 4 vols. (London 1964–66).

less than expert and the territory poorly delineated, the effect is utterly incoherent and not even to be trusted in matters of fact. The effort to achieve fairness and wide coverage in a small compass is wasted when it comes up with statements like 'Tom, Dick and Harry also worked in this field; and so at a later date did Smith, Jones, and Robinson'. The would-be historian of science would be well advised not to take such works as representative of the subject. So far the history of science has been mercifully free from textbooks, though some now seem to be appearing from America. One cannot therefore find in small compass all the names and received opinions that the student ought to know, and the study of the history of science involves the use of scholarly works, articles in journals, and primary sources.

We can begin the constructive part of our survey of histories of science with those written in the past, that is, down to about the end of our period. These might have to be classified under the dubious heading of elderly, not to say obsolete, secondary sources; and there are indeed some which must be placed in this division of limbo. But others were important in their own day, and may still come alive in ours; and there are various reasons why such books can still be very profitably consulted. It may be, first, that they have never, or not yet, been superseded. Thus Robert Small's account of the work of Kepler, published in 1804, has recently been reprinted and remains a standard work, although now it must be supplemented by more recent studies, since nothing on the same scale has been written since; and the same is true of Robert Grant's *History of Physical Astronomy* of 1852.[1] The student of the history of astronomy will also, for the

[1] R. Small, *An Account of the Astronomical Discoveries of Kepler* (reprinted Madison, Wis. 1963); R. Grant, *History of Physical Astronomy* (reprinted New York 1966); J. B. J. Delambre, *Histoire de l'astronomie*, 6 vols. (Paris 1817–27); M. Berthelot, *Les Origines de l'alchimie* (Paris 1885), and *Introduction à l'étude de la chimie des anciens et du moyen age* (Paris 1889); H. Kopp, *Gesichte der Chemie*, 4 vols. (Braunschweig 1843-47) and *Die Alchemie* (Heidelberg 1886); W. Ostwald (ed.), *Klassiker der exacten Wissenschaften* (Leipzig 1889-); T. Birch, *The History of the Royal Society*, 4 vols. (London 1756-57); H. E. Roscoe and A. Harden, *A New View of the Origin of Dalton's Atomic Theory* (London 1896),

same reason, turn to J. B. J. Delambre; while those interested in alchemy and early chemistry will want to consult the works of Marcellin Berthelot, Herman Kopp, and Wilhelm Ostwald. These histories are still valuable because they were meticulously compiled and because there is nothing newer with the same scope; they are often also useful collections of documents. Thus Ostwald's *Klassiker* is a convenient source of chemical papers which to Ostwald about 1900 seemed very important—carefully translated where necessary into German, which is a help when their original language was, for example, Swedish. Similarly Thomas Birch's *History of the Royal Society* is a transcription of the Register Book of the Society and not a history in the ordinary sense; his manuscript sources survive, but sometimes when the originals have disappeared—as with Dalton's papers—the closest that we can get to them is through the writings of somebody who did see them. But it would be unusual to begin the study of the history of the science of a period with detailed books of this kind; the tyro would almost certainly find that they told him more than he wanted to know, and he would like to have some more recent map of the territory to indicate when these old charts should be handled with care.

Most people will therefore begin with modern books on the history of science; but those who have already studied the political, social and economic history of a period and who want to learn about its science might be well advised to turn to histories of science written in or soon after that period. In a sense, these will have been for the most part superseded; that is, replaced, for those seeking detailed information or deep analysis, by more recent works, as happens with all secondary sources. But the picture which the authors give of the objectives of science, and the lists of names of those whom they regard as the greatest of scientists, cannot but interest the historian anxious to look for the problems that faced scientists, and to see what counted as science,

---

reprinted, ed. A. Thackray, (New York 1970); see also A. Thackray, *John Dalton* (Cambridge, Mass. 1972), on the sources which do survive.

rather than to list contributions to the edifice of modern science. In works of this second class, we may for example find considerable space devoted to science in antiquity; classical scholarship has advanced, and better texts are available to us as a rule, but what the historian wants to know is not what Aristotle or Democritos said, or what scholars today believe them to have said, but what use was made of their authority and writings during his period. For this, a history written in that period is a better guide than the best commentary of our own day.

Older histories can thus put us into a tradition which we may find very different from what we might have expected; and we may thus be in a better position to see what kind of questions the scientist would be asking of nature, and what kind of answers he would expect to get. This will be even clearer from those parts of the work in which the historian is describing what was to him the recent past; especially if, as was often the case, he had himself played a part in the events he described, or at least knew those who had. In short, the older historians will tell us more about the milieu in which they were written than about their ostensible subject. Like all books, histories of science have been written with an end in view; they may glorify some nation, religious belief, or philosophical position, and it is the historian's task to determine how far such glorification was typical of the period, trying to distinguish the conventional piety from the remark with a cutting edge. Some works of history, like Sprat's *History of the Royal Society*, could equally well be classified as propaganda; but that does not affect its value as a source for its period.[1] We must be guided in using the older histories by Coleridge's maxim 'until you understand a writer's ignorance, presume yourself ignorant of his understanding'; there is no point in supposing that those respected in their own day were ignoramuses, or that we can see more clearly than they could. It may be interesting

[1] T. Sprat, *The History of the Royal Society* (London 1667); ed. J. I. Cope and H. W. Jones (London 1959); and see M. Purver, *The Royal Society* (London 1967) pp. 9–19. S. T. Coleridge, *Biographia Literaria*, ed. J. Shawcross (Oxford 1907) I, 160.

to demonstrate this in particular cases; but we are not engaged in some sort of competition with our ancestors, but in an attempt to bring to light the assumptions and activities which they took for granted.

The history of science was not seen in our period as distinct from the history of philosophy; as has been said, what we call 'science' was often called 'philosophy' and this usage indicated a genuine close connexion. Some of the best historians of science were philosophers. Thus in Hegel's *Philosophy of Nature* there is a remarkably good account of the science of his youth; and Comte's *Positive Philosophy*, while much more sketchy, did stimulate his disciples to impose his scheme in which religion gave way to metaphysics, to be in turn replaced by the positive knowledge acquired by the methods of empirical science.[1] This historicism attracted among others John Stuart Mill and G. H. Lewes, who wrote an *History of Philosophy* which was intended as an obituary notice for religion and metaphysics; but although the historian of science should look at Mill and Lewes they cannot themselves be described as historians of science. It was Mill's opponent, the polymath William Whewell, who wrote the most interesting history of science in English in the first half of the nineteenth century. He wished to get away from the view that science consisted simply of collections of indubitable facts, and urged that it was only when clear and distinct conceptions were imposed upon reliable data that science could result. This gave him wider sympathies than those attracted to the aridities of Positivism, and his concern with problems and explanations rather than with contributions is apparent; in the manner of the time, he boldly took lessons from the past into the present and argued with the

[1] G. W. F. Hegel, *Philosophy of Nature*, tr. A. V. Miller (London 1970), and tr. M. J. Petry, 3 vols. (London 1971); J. S. Mill, *A System of Logic* (London 1843); A. Comte, *Cours de philosophie positive*, 6 vols. (Paris 1830–42); G. H. Lewes, *The Biographical History of Philosophy*, 4 vols. (1845–46, 2nd ed., 1857); W. Whewell, *History of the Inductive Sciences*, 3 vols. (1837), and *Philosophy of the Inductive Sciences*, 2 vols. (London 1840); see the reviews of reprints of these, *British Journal for the History of Science*, 4 (1969), 399–402.

dead when he considered it appropriate. For general interest, one would today turn to Whewell's writings on philosophy of science rather than on history of science; but to the student of the early Victorian period his history is valuable because it was widely read by scientists and non-scientists, and because Whewell was a learned man in touch with most of the important scientists of his own day.

The same is true of his contemporary Sir John Herschel who became a great pundit on all questions concerned with the sciences, and whose *Preliminary Discourse* was very widely read.[1] It was a treatise on scientific method, making considerable use of historical examples, rather than a formal work on the history or the philosophy of science; but the view it gave of science and its history was very important in its day. Herschel had been one of those responsible for the reinvigoration of mathematics in England in the 1820s, and his appreciation of mathematics broadened what might otherwise have been a 'Baconian' inductivism; this latter spirit informs the history of science written by Baden Powell, another mathematician, making it indigestible but not untypical of its period.

If Whewell and Herschel were the best historians of science writing in English in the first half of the nineteenth century, using history to establish a view of science and a canon of great works, and casting some light on the relationships between the various sciences, the end of the century had its great contribution to the discipline in the electrical engineer J. T. Merz.[2] His plan was a novel one, for he began by describing the state of science in France, Germany and England in the first half of the nineteenth century; and then considered not the histories of specific sciences, but the astronomical, atomic, kinetic, physical, morphological, genetic, vitalistic, psycho-physical, and statistical views of nature, concluding with a chapter on the history of mathematics. This

[1] J. F. W. Herschel, *Preliminary Discourse on the Study of Natural Philosophy* (London 1830); Baden Powell, *History of Natural Philosophy* (London 1837).

[2] J. T. Merz, *A History of European Thought in the 19th Century*, 4 vols. (Edinburgh 1896–1914); the first two volumes deal specifically with science.

approach enabled him to bring out connexions between the sciences, and to handle the whole of the science of the nineteenth century in reasonably brief compass without becoming bogged down in detail. His work brings out the breadth of interest of many scientists of the nineteenth century; combining sound judgement with originality of approach, Merz's is a splendid account of the history of science as it was seen at the end of our period.

By the end of the period, with the writings of Merz, of Pierre Duhem, and of George Sarton, the history of science had emerged as a discipline separate from science;[1] and we have seen that there were nineteenth-century examples of pure rather than applied history of science. For the first part of our period, most of the history was written as part of a work of science, or as biography. While from a work formally on the history of science we can hope to find out what traditions scientists were operating in and what problems faced them, this task should be easier when the scientist is putting his own work into an historical perspective. Scientists have perhaps surprisingly often used historical arguments to justify what might otherwise seem dangerously novel views. This may sometimes take the form of an appeal to authority; thus Dalton and his opponents expounded Newton in support of their view of the nature of chemical combination.[2] Similarly, Thomas Young presented his wave theory of light, at about the same time,

[1] P. Duhem, *The Aim and Structure of Physical Theory* [1914], tr. P. P. Wiener (Princeton, N.J. 1954); G. Sarton, *Introduction to the History of Science*, 5 vols. (Baltimore, Md. 1927–48)—now chiefly valuable for its bibliography.

[2] On Dalton's theory and its reception, see my *Atoms and Elements*, 2nd ed. (London 1970); and D. S. L. Cardwell (ed.), *John Dalton and the Progress of Science* (Manchester 1968); the primary documents are reprinted in *Alembic Club Reprints, 2* and *4* (Edinburgh 1961), and in my *Classical Scientific Papers; Chemistry* (London 1968). T. Young, 'Theory of Light and Colours', *Philosophical Transactions, 92* (1802) 12–48; [H. Brougham], *Edinburgh Review, I* (1803), 450–6. A. Koyré, *From the Closed World to the Infinite Universe* (Baltimore, Md. 1957), ch. II; F. E. Manuel, *Isaac Newton, Historian* (Cambridge 1963), 18. See D. F. Larder's papers on Butlerov, in *Ambix, 18* (1971), 26–48, and *Journal of Chemical Education, 48* (1971), 287–91; the latter includes a translation of Butlerov's paper.

as a development of Newton's ideas, supporting it by copious quotation; while Henry Brougham, the spokesman for his opponents, attacked the theory as an heretical exposition of Newton rather than as bad physics. Both Copernicus and Newton appealed to the shadowy authority of the Pythagoreans for the heliocentric theory and for the theory of universal gravitation respectively; and in nineteenth-century Russia the debate over chemical structure-theory took the form of a controversy upon whether it was in accordance with the main stream of the development of the science.

While on the one hand, revolutionary scientists have often sought to show that their theories are within a long and honourable tradition, and though apparently radical really represent a return to the main stream; on the other hand, opponents of new theories usually begin by attacking them as wild and ill-founded, and then if the theory prevails seek to show that there was nothing new about it. Thus when Lavoisier's theory of combustion and respiration came to be generally accepted in England, patriotic examination of some English chemical writings of the later seventeenth century, particularly those of John Mayow, revealed a belief in 'aerial nitre' which could be regarded as an anticipation of Lavoisier's views.[1] Mayow's writings were indeed republished by Thomas Beddoes, an early convert to Lavoisier's doctrine but no francophile. Darwin's theory met the same fate; at first opponents like Richard Owen, the Duke of Argyll and Bishop Wilberforce made a head-on attack upon it; and when this failed to prevent its general acceptance as a working hypothesis it was urged instead that the theory was no more than a rehash of the views of Lamarck, Erasmus Darwin, scholastic theologians, or even Empedocles. Darwin tried to defend his intellectual property in an introduction to later editions of the *Origin of Species*; and his 'bulldog', T. H. Huxley, did serious historical research to show that the Schoolmen at least had in no sense

[1] T. Beddoes, *Chemical Experiments and Opinions, extracted from a work published in the last century* (Oxford 1790); T. H. Huxley, *Darwiniana* (London 1893), essay V.

anticipated Darwin. Scientists have thus appealed to history probably as much as those engaged in politics, and for much the same reasons.

But there have been some occasions when it has been clear that the main stream of history cannot plausibly be held to have led to the new theory under discussion. In that case the author's appeal to history must take the form of an explanation of why his predecessors, though entitled to respect in general, had failed in this instance to see the light. This use of history can be more interesting than the search for respectable ancestors or for precursors because it can lead to the recognition that the scientists of earlier generations were concerned with different problems. Thus in 1830 Charles Lyell began to publish his *Principles of Geology*, arguing that past changes are to be accounted for in terms of causes operating at the present day.[1] This idea was not completely new, but it had not before been presented with such logic and thoroughness; and the prevailing view, supported by the great authority of Georges Cuvier, was that to explain the changes in fauna and flora revealed by the fossil record periodic convulsions or catastrophies had to be invoked. By a witty and devastating account of the views of his predecessors, Lyell—who had been trained in the law—succeeded in convincing most of his contemporaries that geology had gone wrong because its exponents had been prodigal of violence and parsimonious of time. If geological history were seen as an affair of millions of years rather than hundreds, then earthquakes, floods and volcanoes could produce the changes observed without any need for inexplicable catastrophes. In somewhat similar vein, Ernst Mach examined the history of mechanics to show how the, to him erroneous, doctrines of absolute space and time had got into the science; he did not convert many contemporaries, but he did

[1] C. C. Gillispie, *Genesis and Geology* (Cambridge, Mass. 1951); L. G. Wilson, *Charles Lyell* (New Haven 1972–); M. J. S. Rudwick, 'The Strategy of Lyell's *Principles of Geology*', *Isis, 61* (1970), 4–33; E. Mach, *The Science of Mechanics*, tr. E. J. McCormack (Chicago 1902); A. Einstein and L. Infeld, *The Evolution of Physics* (Cambridge 1938).

convince Einstein who abolished absolute space and time in his theory of relativity, and who himself wrote on the history of science legitimising his views.

We thus find throughout our period scientists appealing in their books and even in their papers to authorities in the history of their science as well as to modern authors more obviously relevant to the discussions in hand. We may also find wider references; the historian who cannot believe, for example, that religion, or even alchemy, is interesting will find more or less of the science of this period opaque. We shall also find historicism among scientists; theories in chemistry, for example, being judged not on their merits in accounting for chemical phenomena alone but also as steps along the route by which chemistry was re-enacting the history of astronomy. Especially about 1800, eminent chemists sought the Kepler or Newton of chemistry, sometimes putting their own names forward; and argued, for instance, that because Newton had shown that all celestial motions were the result of universal gravitation some equally simple law must be found to account for chemical affinity.[1] In the event, the astronomical paradigm proved unprofitable in chemistry; but physical scientists still speak of some discipline being 'at the natural history stage' as though there were some sequence through which all sciences had to pass on their way to respectability. This appeal to historicism must be distinguished from the borrowing of concepts from fashionable sciences; thus the idea of evolution came from biology into chemistry and astronomy in the decades after the *Origin of Species* was published, but this does not necessarily mean that chemists and astronomers thought that their science must follow the historical development of biology.

We have mentioned religion, alchemy, and respectability; and this may remind us that one of the great problems at the beginning of our period was to civilise atomism. In antiquity the atomic theory had been associated with atheism, or at least deism; and Harriot and Hobbes had by the middle of the seventeenth century reinforced this association. Atomism, and the materialism to

[1] A. Thackray, *Atoms and Powers* (Cambridge, Mass. 1970).

which it led, could not but lead to civil disorder or tyranny, however valuable it might seem to be to the man of science.[1] The energy of Gassendi, of Robert Boyle, and of Henry More was directed to the reconciling of the atomic theory with Christianity; and Ralph Cudworth, in his *True Intellectual System* of 1678, crowned this reconciliation with a new pedigree for atomism, which had first been expounded by Moses and had been merely corrupted by Democritos, Epicurus and Lucretius. The atoms, being composed of inert brute matter, could not move or stick together, let alone think; and thus immaterial substance was necessarily required. Since atoms could not be destroyed, no more could immaterial substance; acceptance of atomism therefore entailed belief in the immortality of the soul. Since the atoms were all composed of the same stuff, namely matter, and constituted lead or gold only because of their arrangement, alchemy was a possibility; and Boyle was responsible for the repeal of the old statute against it. Nearly a century after Cudworth, the Unitarian materialist Joseph Priestley used a different historical argument, that the immortality of the soul was a Platonic corruption not present in primitive Christianity, to support his view that matter was not passive and inert, but active and composed of centres of force; and with his support for the French Revolution proved to his contemporaries the connexion between materialism and civil disorder.

Natural historians and navigators also took an interest in history, but of a rather different kind. Particularly at the beginning of our period, natural history was as much a matter of erudition as of observation; for before the concise but full

[1] R. H. Kargon, *Atomism in England from Harriot to Newton* (Oxford 1966); P. Heimann, 'Nature is a perpetual worker', *Ambix*, *20* (1973), 1–25; J. E. McGuire, 'Boyle's Conception of Nature', *Journal of the History of Ideas*, *33* (1972), 523–42; my paper, 'Chemistry, Physiology, and Materialism in the Romantic Period', *Durham University Journal*, *64* (1972), 139–45; J. B. Morrell, 'Professors Robison and Playfair, and the *Theophobia Gallica*', *Notes and Records of the Royal Society*, *26* (1971), 43–63; R. E. Schofield, *Mechanism and Materialism* (Princeton 1970); C. B. Brush (ed. & tr.) *Selected Works of Gassend*, (New York 1972).

descriptions of Linnaeus and his disciples were introduced in the middle of the eighteenth century it was not easy to know whether an animal or plant had been described before or not, and a critical study of the literature was necessary. This was in line too with the Renaissance interest in identifying the plants and animals described by Aristotle, Theophrastos, Dioscorides and Pliny. It was also important to look for records of creatures being seen, or being observed to breed, in a given locality; and standard works of the eighteenth and nineteenth centuries are surprisingly full of records old and new. Sometimes these are uncritical; as Dürer's rhinoceros drove out from illustrated works more accurate pictures of the animal, so anecdotes were passed on, and (for example) Linnaeus and Thomas Pennant gave currency to the story that squirrels crossed water on a piece of bark as a raft, using their tails as a sail.[1] Similarly, before the coming about 1770 of accurate means of fixing longitude by the chronometer or by lunar observations, it was difficult to know whether or not one had arrived at, for example, the fabulously rich Islands of Solomon described by a Spanish navigator many years before; geography required a critical examination of sources just as natural history did. We can discover what was seen as the main line of development in these various sciences from the authorities cited in works of this kind; and indeed natural history preserves this historical interest with its conventions on nomenclature, and its use of 'type' specimens where the specimen from which a species was first described—or sometimes only a drawing of it—is kept as the touchstone by which it can be seen whether some other creature belongs to the same species or not.

We remarked before that historians may be most valuable when writing about the science of their own day; and we do find that numerous scientists have given us accounts of events in which they, or those they know, played a part. Thus Thomas Thomson, the author of the best-selling chemistry textbook in

[1] W. George, *Animals and Maps* (London 1969); T. Pennant, *British Zoology* (London 1776), I, 96; *Byron's Journal of his Circumnavigation 1774–1776*, ed. R. E. Gallagher (Cambridge 1964) (Hakluyt Society).

which Dalton's atomic theory was first announced, and editor of the journal in which J. J. Berzelius published his chemical symbols, wrote a *History of Chemistry* presenting an insider's view of the subject during his professional lifetime.[1] Thomson's journal, *Annals of Philosophy*, carried each year a review of the progress of chemistry, and Berzelius wrote even more famous and useful annual reports; but these can hardly count as history, though they are valuable for the historian today. Reviews of a rather longer period were published by Delambre and Cuvier in 1810; these were reports delivered to Napoleon on the progress of science since 1789. Delambre took astronomy, physics and mathematics, and Cuvier chemistry, biology and geology; the addresses are larded with flattery to the Emperor and are nationalistic in tone, but not less valuable and interesting for that.

Cuvier, like Fontenelle, Condorcet, and François Arago was famous for the *Eloges* that he prepared for deceased Academicians; and in the nineteenth century the idea became general that scientific societies should publish in their journals obituaries of prominent members who had died. These, written by a colleague, disciple or rival of the dead man, are a very valuable source of biographical information, though often more discreet than one would wish; for a clear statement of a man's career in science as it appeared to contemporaries such obituaries are often much more valuable than those two—or three—volume official biographies, the fruits of filial piety in the nineteenth century.[2] Such works should not be despised, though there are few that one would read simply for pleasure, and though editors of that date were often, by our standards, cavalier in their publishing of manuscripts, excising passages without giving any indication that they had done so, for example. Biographies of scientists were not

[1] T. Thomson, *The History of Chemistry* (London 1830–31); J. B. J. Delambre, *Rapport Historique sur les progrès des sciences mathématiques depuis 1789* (Paris 1810); G. Cuvier, *Rapport Historique sur les progrès des sciences naturelles depuis 1789* (Paris 1810).

[2] See B. Z. Jones (ed.), *The Golden Age of Science* (New York 1966); H. Hartley, *Humphry Davy* (London 1966), 154; H. Mayhew, *The Wonders of Science or Young Humphry Davy* (London 1855), is not on Hartley's list.

always written to exalt the memory of their subject: Paris' *Life of Sir Humphry Davy*, for example, was written in consultation with Lady Davy to belittle him by under-rating his scientific attainments and deriding his social pretensions; but this soon produced a counter-biography written by Davy's brother John. Scientists and engineers who, like Davy, rose to eminence from humble origins attracted Victorian biographies; Henry Mayhew wrote an edifying *Life* of Davy which has no claims whatever to accuracy, but the most famous author in this genre, Samuel Smiles, was careful to verify his references, and his biographies are valuable documents.

For the nineteenth century, then, we can generally find obituaries and full-length biographies of major figures in the history of science, though these may well not tell us what we want to know. For the earlier part of our period, there are fewer of these contemporary sources; and throughout the period we may be frustrated by finding that someone of interest to us as a scientist or patron of science did not appear to his biographer or even to himself as such. Thus Bishops Seth Ward and Samuel Wilberforce were more involved with science than their biographies suggest, and Joseph Priestley in his *Autobiography* devotes much more time to his theological controversies and clerical activities than to his chemical researches. For French scientists of note in the eighteenth century, and for Foreign Members of the Académie des Sciences, we have the *éloges*; for others, we have the useful compilation by Benjamin Martin, the Newtonian populariser and instrument-maker; the *Biographia Britannica*, a very valuable source of usually reliable information; the *Harleian Obituaries*, which are telegraphic but can be a useful guide to fuller sources; and the obituaries which appeared in *The Gentleman's Magazine*. The eighteenth and nineteenth centuries were the great age of the encyclopedia; encyclopedias were usually strongest on science and technology, and often include biographies of scientists. At the beginning of the nineteenth century, an abridgement of the *Philosophical Transactions* of the Royal Society was prepared, and this includes some brief biographies of men of science of the seventeenth and eighteenth centuries. For those who were professors at Gresham College in London, we have Ward's collection of *Lives*; and for Oxford men of the seventeenth century there is Anthony à Wood's *Athenae Oxonienses*, which, like Aubrey's *Brief Lives*, tells us about a number of natural philosophers. Less entertaining histories of colleges are often useful sources of information about their alumni; and anecdotes about men of science may be found in such collections as those of Joseph Spence and Isaac d'Israeli.[1]

These anecdotes may be *bien trouvê* rather than wholly reliable; stories about Newton—that he found Euclid obvious and Descartes nonsense, and that he could not tot up a column of figures—circulated in the eighteenth century and can all from manuscript sources be shown not to be true. An author should not be believed merely because he happened to be a contemporary or near-contemporary of the events he described: as we know, stories improve in the telling and the historian must try to check them; or if he cannot he must repeat them as second-hand, and as telling us as much for example about Spence or Voltaire as about Newton. Sometimes contemporary accounts may have been designed to mislead; thus a circumstantial story of the death and funeral in England of the chemist and projector J. J. Becher, by an associate, seems to be a fiction put about probably to put Becher's numerous creditors off his track.[2]

Autobiographies are equally liable to err; old men forget, and nobody is on oath in his memoirs. When an historian, in possession of contemporary documents, recently interviewed participants in a scientific controversy of some forty years before, their recollection of what happened diverged as might be expected

[1] B. Martin, *Biographia Philosophica* (London 1764); *Biographia Britannica*, 6 vols. (London 1747–66); W. Musgrave, *Obituaries prior to 1800*, ed. G. J. Armytage, 5 vols. (London 1899–1901); *Gentleman's Magazine*, *1* (1731)–*303* (1907); *Philosophical Transactions*, ed. C. Hutton *et al.*, 19 vols. (London 1809); J. Ward, *Lives of the Professors of Gresham College* (London 1740); A. à Wood, *Athenae Oxonienses*, 2 vols. (London 1691–92); J. Aubrey, *Brief Lives*, ed. O. L. Dick (London 1949); J. Spence, *Anecdotes*, ed. B. Dobree (Fontwell 1964); I. D'Israeli, *Curiosities of Literature*, ed. E. Bleiler (New York 1964).

[2] Becher's 'death' is being investigated by W. A. Campbell.

from what the documents revealed.[1] In retrospect their views seemed tidier and closer to those now held than was really the case; and the order of events was also made more logical. Old men may mellow or become embittered, and their account of the acrimony of the debates of their youth cannot be implicitly trusted; there are fashions in polemic, too, and what is plain speaking in one generation may seem intolerable rudeness in another, and vice versa; it is difficult to judge the intensity of feeling in discussions of long ago, and to remember exactly what were the points at issue. While scientists in their memoirs—and in biographies written by their friends—tend to appear more logical and less passionate than they really were, there have been some who have recounted their mistakes and their fortunate accidents with as much gusto as their successful inductions and deductions. Examples of this are Kepler and Joseph Priestley, who incorporated accounts of their errors and serendipities into their scientific works.[2] Priestley maintained that most scientists covered their tracks, not making discoveries in the logical way in which they wrote them up, but stumbling upon them; but his own reputation, then and since, suffered because of his candour. Priestley need no more be believed, naturally, when he reports stumbling on something by accident than some contemporary who brings forward a long chain of reasoning which led him directly to his goal; the participant's evidence is one factor which the historian must take into account, but it is not something that settles a question beyond doubt.

Priestley's democratic and Unitarian opinions were another factor in the suspicion with which he was regarded by contemporary scientists, and these are indeed emphasised in his autobiography. An earlier Unitarian scientist was William Whiston, Newton's successor at Cambridge, who was deprived of his

[1] This exercise was carried out by Dr M. A. Hoskin.

[2] On Priestley, see e.g. J. Davy, *Memoirs of Sir H. Davy* (London 1836), I, 223; R. E. Schofield, *A Scientific Autobiography of Joseph Priestley* (Cambridge, Mass. 1966), 268; J. Priestley, *Autobiography*, ed. J. Lindsay (Bath 1970), 94, 125. On Kepler, A. Koestler, *The Sleepwalkers* (London 1959).

Chair because of the zeal with which he propagated his unorthodox opinions; in his racy autobiography, and his *Life* of Samuel Clarke, a divine also more orthodox in his Newtonianism than in his theological views, he gives us a valuable picture of science in its relation to religion in the early eighteenth century.[1] Autobiographies were not infrequently left in manuscript to be published by the heirs of the writer, and many Victorian *Lives and Letters* contain a brief autobiography, or a fragment of one. Whether short or long, such documents were edited, and passages felt to be offensive or undignified may have been excised. Darwin's autobiography is an example; a full edition of this has now been published, and comparison of the two texts enables us not only to see what Darwin actually wrote but to assess the mores of his family. Autobiographies like Darwin's were often written, ostensibly at least, not with publication in mind but for family records; this should be borne in mind in assessing what does and does not appear in them.

For recently dead scientists, the *Dictionary of National Biography* and the works that correspond to it for other nations, are valuable if uneven sources;[2] for those who died long before the dictionary was compiled, the entry must be read with all the suspicion due to an elderly secondary source, howbeit a very useful one. There are recent biographical dictionaries of scientists available, which are very valuable compilations for the historian, referring him to primary and to recent secondary literature; this is especially true of the *Dictionary of Scientific Biography*, which is currently appearing. These, like modern full-length biographies, do not replace contemporary accounts and must not be regarded as infallible; but they can provide an excellent lead into the subject,

[1] W. Whiston, *Memoirs*, 3 pts. (London 1749–50); *Historical Memoirs of the life of Dr. Samuel Clarke* (London 1730). C. Darwin, *Autobiography*, ed. N. Barlow (London 1958); on biographies generally, see my *Natural Science Books in English* (London 1972); this also lists bibliographies.

[2] *Dictionary of National Biography*, 63 vols. (1885–1900); *Australian Dictionary of Biography* (Melbourne 1966–); *Dictionary of Scientific Biography* (New York 1970–); *T. I. Williams* (ed.), *Biographical Dictionary of Scientists* (London 1969); A. Debus (ed.), *World Who's Who in Science* (Chicago 1968).

and save a great deal of time. It is a general feature of such dictionaries that many of those in whom one is most interested are left out, and that most space is as a rule devoted to those about whom information can easily be found anyway: they are invaluable for their brief biographies of those peripheral to one's main concern, so that one can avoid major blunders; and for their bibliographies. The same applies to modern biographies, which can be a very valuable introduction to a period.

Apart from the brief bibliographies in the biographical dictionaries, and in biographies some of which do contain reasonably full bibliographies, there exist separate and detailed bibliographies of writings by and about various important men of science. Such books can save the historian a great deal of time in hunting through indexes of journals and library catalogues, and are invaluable. The bibliography of one man may also cast light upon the writings of contemporaries; a man will have reviewed the works of another, or written for or against him. And it is interesting to know where and when a man's writings were translated, which papers or books were translated, how full the translation was, and whether it contained new material. These facts give us some data upon which to base judgements about contacts between scientists in different countries, complementing the biographical information on who was elected a foreign member of which societies, and at what period of his career.

Where we do not have these individual bibliographies, for the writings of scientists of the nineteenth century we can turn to the *Royal Society Catalogue of Scientific Papers*.[1] The subject-index for

[1] *Royal Society Catalogue of Scientific Papers, 1800–1900*, 19 vols. (London 1867–1925). Critical bibliographies are published every year in *Isis*; in addition to the British Museum Catalogue, it is worth consulting that of the British Museum (Natural History), which is also published. See for example also J. Ferguson, *Bibliotheca Chemica*, 2 vols. (Glasgow 1906); R. B. Webber and H. P. Macomber, *A Descriptive Catalogue of the Grace K. Babson Collection of the Works of Sir Isaac Newton*, 2 vols. (New York 1950–55); and in general, see J. L. Thornton and R. I. J. Tully, *Scientific Books, Libraries, and Collectors*, 3rd ed. (London 1971); A. J. Walford, *Guide to Reference Material*, 3 vols., 2nd ed. (London 1966–70).

this enormous international project was never completed after it came to a halt in the First World War; but the name-index volumes are invaluable to the historian. Here again we can find information about translations, and about papers which appeared in more than one journal. Such multiple publishing was a feature of nineteenth-century science; some bodies, as for example the Royal Society, insisted that papers sent to them must appear first in their journal, so that one had to choose between prestigious publication there or speedy appearance elsewhere. The more popular journals often filled a lot of space by reprinting, often in parts, important papers from more exalted but less generally accessible publications. The *Royal Society Catalogue* usually indicates where this has happened; but its range does not include the humblest journals, addressed to 'mechanics', and for papers in these journals there is no alternative but to search through. This drudgery is not infrequently rewarded by the discovery of papers on unexpected subjects, and of entertaining controversies that carry on through several issues; but such searching is for the most part rather tedious, since having some end in view one cannot linger over the curious contributions and delightful juxtapositions. As a guide to secondary literature in the history of science, there are the *Isis* bibliographies published each year. *The Royal Society Catalogue* does not of course list books; for books of this as of earlier periods, one must consult the catalogues of great libraries such as those of the British Museum and Bibliothèque National, the *American Union Catalog*, university library catalogues, and catalogues of scientific societies or institutions. When these catalogues are published the historian's life is made much easier. There are in addition various bibliographies of scientific books, which are often indeed catalogues of some particular collection.

Biographies and bibliographies are valuable to the historian but they are not themselves history, which, whatever some of our ancestors may have supposed, is more than a collection of biographies or of gleanings from a library. We shall therefore turn last to what most of those contemplating work in the history of science will have turned to first: modern works on the history of science. Here the first part of our period is on the whole more written about than the second; nineteenth-century science is a vast and complicated field for which we have some itineraries and maps of a few localities but no general survey as yet. Historians have followed, as one might expect, different lines of approach; works of major importance in the last generation have taken the form of biographies, of new editions of letters, lectures or books, of works of intellectual history, both narrative and in essays, of institutional history, and of works relating the history of science to that of literature, fine art and techniques, and to political and economic history.

The majority of biographies tell us more about their subject than we already knew, but particularly if they are brief they may not do much more than that. But there are some which force upon historians a reconsideration of the received view of a period and give us a new vision of what constituted scientific activity at some time in the past. Such books thus stimulate new work instead of summing up work already done. Thus the recent *Lives* of William Harvey, by Geoffrey Keynes and particularly by Walter Pagel, make us look afresh at his time, when the clear division between magic and empirical science had not been made. Maynard Keynes' brief account of Newton as the last of the magi had pointed in the same direction; and the recent *Life* of John Dee seeks to show him not merely as magician but as a reputable figure in the prehistory of modern science. Recent biographies of Linnaeus have brought to life a man who had seemed a dry classifier, and have explored the basis of his taxonomy, which made natural history the leading science of the mid-eighteenth century; and at last we have had a lengthy study of Boerhaave, whose importance in the history of chemistry is once more apparent. Pearce Williams' *Faraday* showed its subject as a much more theoretical and metaphysical scientist than he had generally been taken to be and called attention to little-regarded traditions within the chemistry and physics of the nineteenth century; and Meadows' *Lockyer* provides an excellent introduction to the small world of science in England about 1900. Writings about Darwin have been legion; and of Cuvier there is now a sympathetic study.[1]

These biographies cover the whole life of their subject; but sometimes better illumination is achieved when only a brief period of a man's life is closely examined.[2] Thus Guerlac changed our view of Lavoisier's gradual appreciation of the role of air in combustion in his *Lavoisier—the Crucial Year*; and Gweneth Whitteridge showed the value of concentrating upon Harvey's discovery of the circulation of the blood. Other authors have constructed a biography of their subject by a selection of his own writings with linking commentary;[3] thus Schofield has given us a new portrait of Priestley as a scientist, and Lysaght has succeeded in showing that Sir Joseph Banks must be taken much more seriously as a man of science than he has often been in the past. We have also recently had an edition of Banks' *Journal* kept on his voyage to Tahiti and New South Wales with Cook, edited by Beaglehole who has also given us a splendid edition of Cook's *Journals.* Bodies like the Hakluyt Society and the Navy Records Society produce a splendid series of editions of journals and documents connected with voyages and exploration, some of which were of considerable importance as scientific

[1] G. Keynes, *William Harvey* (Oxford 1966); W. Pagel, *William Harvey', Biological Ideas* (Basel 1967); Royal Society, *Newton Tercentenary Celebrationss* (Cambridge 1947), 27–34, and F. E. Manuel, *A Portrait of Isaac Newton* (Cambridge, Mass. 1969); P. J. French, *John Dee* (London 1972); W. Blunt, *The Compleat Naturalist* (London 1971), and F. A. Stafleu, *Linnaeus and the Linneans* (Utrecht 1971); G. A. Lindeboom, *Herman Boerhaave* (London 1968); L. P. Williams, *Michael Faraday* (London 1965); A. J. Meadows, *Science and Controversy* (London 1972); P. J. Vorzimmer, *Charles Darwin: the Years of Controversy* (London 1972); W. Coleman, *Georges Cuvier Zoologist* (Cambridge, Mass. 1964).

[2] H. Guerlac, *Lavoisier—the Crucial Year* (Ithaca, New York 1961); G. Whitteridge, *William Harvey and the Circulation of Blood* (London 1971).

[3] See p. 57, note 2; A. M. Lysaght, *Joseph Banks . . .* (London 1971); *The Endeavour Journal of Joseph Banks*, ed. J. C. Beaglehole, 2 vols. (Sydney 1962); *The Journals of Captain James Cook*, ed. J. C. Beaglehole, 3 vols. in 4 (Cambridge 1955–67); A. Day, *The Admiralty Hydrographic Service 1795–1919* (London 1967); M. Deacon, *Scientists and the Sea, 1650–1900* (London 1971).

undertakings; and we do now have brief general histories of hydrography and oceanography.

We have also recent editions of the correspondence and papers of Henry Ashmole, of Newton, of Henry Oldenburg the first secretary of the Royal Society, of Tobern Bergman, of Joseph Henry, and of Michael Faraday, all of which indicate how contact between men of science was maintained at different periods; and with the publication of Darwin's notebooks we can see more of how his mind worked. Gunther's series *Early Science in Oxford* is a very useful collection of letters, papers and books chiefly of the seventeenth century; and there are other useful collections of documents which can be of great value to the historian. Recent annotated translations include one of the curious protogeological *Telliamed*; and much more important, two translations of Hegel's *Philosophy of Nature*, one of which is fully annotated and cannot but modify our view of science at the beginning of the nineteenth century. The coming of photolithography has meant that a great many reprints of important books in the history of science are now available; some of these have helpful introductions. Gillispie's account of Lazare Carnot as a scientist makes it easier to see how his son Sadi approached the problems of steam-engines as he did on his way towards the second law of thermodynamics.[1]

There is much to be said for plunging in among such attractively—one hopes not seductively—edited primary sources without lingering too long among the secondary materials; but one does need some background, and it is more than useful to have new perspectives indicated. The most stimulating secondary sources often indeed do not deal with the field or period under

[1] C. H. Josten (ed.), *Elias Ashmole*, 5 vols. (Oxford 1966); I. Newton, *Correspondence*, ed. H. W. Turnbull *et al.* (Cambridge 1959–); H. Oldenburg, *Correspondence*, ed. A. R. and M. B. Hall (Madison, Wis. 1965–); T. Bergman, *Foreign Correspondence*, ed. G. Carlid and J. Nordström (Stockholm 1965–); J. Henry, *Papers*, ed. N. Reingold (Washington, D.C. 1972–); M. Faraday, *Selected Correspondence*, ed. L. P. Williams, 2 vols. (Cambridge 1971); B. de Maillet, *Telliamed*, ed. and tr. A. V. Carozzi (Urbana, Ill. 1968); see note 1, p. 46 for the Hegel; C. C. Gillispie, *Lazare Carnot, Savant* (Princeton 1971); R. W. T. Gunther, *Early Science in Oxford*, 14 vols. (Oxford 1923–54).

consideration at all, but simply indicate the kind of questions to ask by demonstrating what interesting answers such questions provoke elsewhere. Koyré, in books and essays, showed the value of treating the history of science as intellectual history;[1] in recent years this has proved a most fruitful line to follow, and it is with works of this kind that most people will begin their reading in the history of science. Such works need not be confined to internal history; indeed much work has been done on the interactions between science and philosophy right through our period; and they may or may not be whiggish in approach. But except in the hands of a master, intellectual history is apt to seem bloodless; for even the most intellectual of scientists do not spend all their time thinking. Much science is concerned with more mundane matters of manipulation and administration, and those financing science have usually been more interested in possible applications than in metaphysical doctrines or elegant theorising. Considerable effort has therefore gone into studies of scientific institutions, and these studies can complement the works of intellectual history.

This emphasis is not new; the Royal Society and the Académie des Sciences had their first historians when they were still in their youth, though perhaps these histories should be described as apologies. In the nineteenth century,[2] Becker wrote a gossipy account of the scientific institutions in London, and Sir John Barrow an anecdotal account of the Royal Society which took the form of biographical sketches of prominent members he had known during his forty years of membership. More serious studies were made of the Royal Institution, and of the Literary and Philosophical Society of Newcastle-upon-Tyne; in the

[1] A. Koyré, *Metaphysics and Measurement* (London 1968); *The Astronomical Revolution* (London 1973).

[2] B. H. Becker, *Scientific London* (London 1874); J. Barrow, *Sketches of the Royal Society* (London 1849); H. B. Jones, *The Royal Institution* (London 1871); R. S. Watson, *The Literary and Philosophical Society of Newcastle* (London 1897); D. S. L. Cardwell (ed.), *John Dalton and the Progress of Science* (Manchester 1968)—A. Thackray is studying the Literary and Philosophical Society; K. T. Hoppen, *The Common Scientist in the 17th Century* (London 1970), describes the Dublin Society.

former study, by Henry Bence Jones, the Institution—the most important centre of research in physics and chemistry in nineteenth-century London—is overshadowed by the great men who worked and lectured there, but the latter does give one a fair idea of a provincial scientific society of no great distinction. There is a recent study, again valuable for establishing a norm, of the Dublin Philosophical Society which flourished at the end of the seventeenth century; and the Manchester Literary and Philosophical Society has been often written about, but chiefly by adulators of Dalton so that up to the present we have little idea of what generally went on, or who belonged to it and why.

Of the more exalted institutions, the Royal Institution is now being studied as a phenomenon in its own right and not merely as the *locus* of Young, Davy, Faraday, Tyndall, Dewar and Bragg; and it has been suggested that it should be seen as the creation of improving landlords who got what they hoped for in Davy's lectures on agricultural chemistry and researches on tanning. Other scientific societies which have been recently studied include the Académie des Sciences in Paris, the Academie del Cimento in Florence, and the less formal Societé d'Arcueil which flourished in Napoleonic France under the patronage of Laplace and Bethollet. About the ancestry and founding of the Royal Society—which has had several historians—there has been considerable controversy in recent years; as there has over the connected question of the social background and religious affiliation of scientists in seventeenth-century England.[1]

Science and literature have had a close relationship throughout our period, which has not perhaps been sufficiently studied

[1] See M. Berman, 'The Early Years of the Royal Institution', *Science Studies, 2* (1972), 205–40; a full-scale history of the Royal Institution is planned, under the editorship of Frank Greenaway. R. Hahn, *The Anatomy of a Scientific Institution; the Paris Academy of Sciences, 1666–1803* (Berkeley 1971); W. E. K. Middleton, *The Experimenters: a Study of the Accademia del Cimento* (Baltimore, Md. 1972); M. P. Crosland, *The Society of Arcueil* (London 1967); M. Purver, *The Royal Society* (London 1967); the review by M. B. Hall in *History of Science, 5* (1966), 62–76; and the various papers in *Notes and Records of the Royal Society, 23* (1968).

although Marjorie Nicolson has pointed a way.[1] Particularly for the science of the Renaissance, the art-historical approach of Frances Yates has proved valuable; for the later period we have the splendid work of Francis Klingender, but more studies on the iconography of science would be valuable, as well as histories of psychological optics. We have accounts of such science-based processes as early lithography and photography, and a guide to what pigments artists used. To connect science and the fine arts demands a light touch; with the useful arts the connexions are closer, particularly by the nineteenth century, and both the technology based upon science and the science dependent upon technology have been written about. As yet, despite the work for example of Christopher Hill on the early years of the seventeenth century, little enough is known of the relations of the history of science to political history; this is a region in which historians of science are likely to meet more ordinary historians, and it is to be hoped that if they do they will not be so concerned with the intellect and with the internal history of science that they will have nothing to say to them. But we, concerned here with sources and their use, need linger no longer among these secondary materials but can pass on to see what manuscript sources are available and what can be done with them.

[1] M. Nicolson, *Newton demands the Muse* (Princeton 1946), and *Science and Imagination* (Ithaca 1956); F. Yates, *Giordano Bruno* (London 1964); F. Klingender, *Art and the Industrial Revolution*, ed. A. Elton (London 1968); M. Twyman, *Lithography 1800–1850* (London 1970); R. D. Harley, *Artists' Pigments 1600–1835* (London 1970); W. Blunt, *The Art of Botanical Illustrations* (London 1950); C. Hill, *Intellectual Origins of the English Revolution* (Oxford 1965); V. Ronchi, *The Nature of Light* (London 1970); A. I. Sabra, *Theories of Light from Descartes to Newton* (London 1967).

CHAPTER 3

Manuscripts

In our field as in others, manuscripts have the priority over other written sources. An experiment is described—one hopes—in a laboratory notebook before being published, and a paper is written out, perhaps in several drafts, and revised before it appears in print. Very often a man will be more open about what he is doing in a letter or a diary than in a book or a formal lecture; and for lectures, notes used by the lecturer or taken by someone in the audience will often give a closer idea of what was said than will a published version polished up some time after the event. And if we are to get behind the scenes, and to find out what went on below the surface in scientific societies or institutions, or in colleges and universities, then we must rely in various degrees upon minutes of meetings, letters, reminiscences, and so on. When we look at natural history or exploration, we find similarly that the notes made in the field and the rough sketches from life or from a specimen only recently dead are prior to, and may well be more accurate than, the polished and handsomely illustrated published work to which they led. Indeed sometimes it may be that without the notes we cannot tell exactly what a naturalist saw or where he went.

Despite this priority, the historian would generally be ill-advised to begin with the manuscripts. The time to turn to manuscript sources comes when the printed material available has been mastered. There is little point in spending hours deciphering a manuscript that was published unless one is familiar with the published version and is looking for differences; and scientific letters, and notebooks or commonplace books, cannot be properly understood until the historian has read the publications familiar to his authors. It is from published materials that the 'norm' is best established; from them semantic changes are best followed;

and they it was that the majority of a man's contemporaries—and everybody in the next generation—actually saw. This applies particularly when we are doing intellectual history; if we are engaged upon institutional history there may be little choice but to plunge straight in among the manuscripts, and the same may apply to the history of technology. But even in these spheres time spent in reconnaisance, that is in reading published primary and secondary materials at least for background, will not be wasted. Historians should aim at broad as well as deep knowledge; particularly in the history of science and technology those editing or interpreting documents have often misjudged them because they have known too little of the customs and commonplaces of the day.[1] This can even be serious; thus in a nineteenth-century edition of the letters of the eminent biologist John Ray we find the editor led into confusion by his ignorance of the old and new style of dating. In seventeenth-century England the year began in March, and the period from January to March of what we would call 1672, for example, would have been written 1671 or perhaps 167½. The Julian calendar was also still in use, so that English dates were behind Continental ones. Odd though it may seem, therefore, a letter from England sent on 1 February 1671 could be a reply to one sent from France on 2 February 1672; if an editor is not prepared for this he may make nonsense of a correspondence. Muddles even similarly arise if one is not aware that eighteenth-century French navigators based their longitudes upon Paris rather than Greenwich; that units such as miles and feet meant

[1] On the general problems of editing correspondence, see J. Sutherland and G. Haight, *T.L.S.* (26 Jan. 1973), 79–80, 87–89. C. E. Raven, *John Ray, Naturalist* (Cambridge 1942), xiv. On units, see M. P. Crosland, 'Nature and measurement in 18th-century France', *Studies on Voltaire and the eighteenth century, 87* (1972), 277–309; on units used, see *Annals of Philosophy, 1* (1813) 452–7; for a curious 'backwards' scale of temperature, see S. P. Krashnennikov, *The History of Kamschatka*, tr. J. Grieve (London 1764), 62, 74; on ship's time, see *The Journals of Captain James Cook*, ed. J. C. Beaglehole, 3 vols. in 4 (Cambridge 1955–67), I, cclxxxvi; II, cxli. R. E. Zupko, *A Dictionary of English Weights and Measures from Anglo-Saxon Times to the 19th century* (Madison, Wis. 1968); A. R. Clarke, *Standards of Length* (London 1866).

rather different things in different places; and that ship time, astronomers' time, and civil time also differed so that the same event may be differently and yet correctly dated.

Given then that the relevant published material has been read, how do we find the manuscript sources that will help us to answer questions which the printed sources leave open? The *Dictionary of Scientific Biography* indicates where archives of the various scientists may be found, if they were known to the person writing the biography; and recently a guide to the archives of British men of science has appeared in microform, edited by Roy MacLeod.[1] These are invaluable guides; and other biographical dictionaries may give indications of where a man's papers may be found. Where a bibliography of a man's writings has been published, it should naturally be consulted and it should describe the manuscripts. The publication of a bibliography usually results in the appearance of previously unknown letters or notebooks; and letters of eminent scientists come up regularly in sales—though it is often hard to find out who bought them—and are published in journals; here the *Isis* bibliography is a help. But there must still be quantities of papers in the attics of descendants of men of science, who may be sought out, with luck, in genealogical works. The bulk of a man's papers may well be found at the college, university or institution where he worked. Thus Davy's letters and notebooks are mostly at the Royal Institution, and those of the Becquerel family at the Académie des Sciences in Paris. Many of Faraday's papers are at the Royal Institution, where he worked; and there is also, appropriately enough, a substantial collection at the Institution of Electrical Engineers. Great libraries have often accumulated, by gift or purchase, collections of archives of men of science; thus in the British Museum library there are further letters of Davy: and scientific societies may retain

[1] *Dictionary of Scientific Biography* (New York 1970–); R. M. MacLeod (ed.), *Archives of British Men of Science* (London 1972); the *Isis* cumulative bibliography comes out in annual parts, and that for 1913–65 in 2 vols., ed. M. Whitrow (London 1971). On sources for physical science in the early nineteenth century, see L. P. Williams in *History of Science*, *1* (1962), 1–15.

manuscripts of papers read to them or addresses delivered on anniversaries; thus the Royal Society has manuscripts of papers and addresses by Davy.

When one can find the manuscripts of scientists of the nineteenth century, one may find a daunting quantity; and indeed this may be true also of men of science from the earlier part of our period. In order to know what to look for we must, as previously noted, be acquainted with published material. Before the days of photocopies and xerox copies, people often copied chunks of published material into their notebooks; manuscripts therefore are not always original material.[1] Much of the archive material may be trivial, and much may be very hard to read; small writing, done with a spluttering pen, may fill the page, and we may find that when the page was filled the writer turned it on its side and filled it again going across the first writing. Unless we begin with a fair idea of what such notes or letters are likely to be about we shall have a great deal of difficulty in deciphering them; and when we only have a file copy, done by that curious Victorian offset process on to damp paper, our problems will be greater still. Official letters and minutes beautifully written out by clerks are a different matter; they are easy to read but have not the immediacy and unguarded nature of holograph manuscripts. The documents closest to the scientific work of those we are studying are the notebooks kept in the study, the laboratory or the field, which should show us the man's mind, hand and eye at work. They vary considerably in character; thus at the Royal Institution there are the beautifully organised volumes of Faraday's laboratory notebooks, with the paragraphs numbered for cross-referencing, as well as the untidy sheets of Davy's notes, collected where possible and bound up by Faraday. The nature of the notes confirms what we read in contemporary accounts about the way the two men worked. In his research—though not in his lecture demonstrations—Davy was an impulsive

[1] This is true, for example, of the MSS of Thomas Wright at Durham and Newcastle; see the introduction by M. A. Hoskin to the reprint of Wright's *Original Theory (1750)* (London 1971).

and rapid experimenter, who bewildered onlookers, took no safety precautions, and ruthlessly turned apparatus to new uses, while Faraday was much more careful and methodical. Indeed Davy's work was described by his great and systematic contemporary Berzelius as brilliant fragments; in this case the notebooks are in accord with the general character of his scientific research, but it will not always be so.[1] Berzelius himself admired the untidiness of Davy's laboratory, remarking—and the sentiment can, *mutatis mutandis*, encourage us all—that a tidy laboratory was a sign of an idle chemist; but it could well be that to fit in with accepted usages an impulsive, opportunistic, and intuitive author will write up his work in polished and deductive form, or that a cautious soul will present the conclusions of a sound induction as the result of happy accident or cosmic disclosure. The laboratory notebooks may help us to see whether or not this has happened.

They also help us to date experiments. This may matter if we want to establish a priority; although X's account was not published until after Y's, he may well have done whatever it was first.[2] Priority is not perhaps very interesting, but we may well be able to confirm that X's and Y's experiments were done independently. If the notebook is dated—and correctly dated—we may be able to establish that the work was done before one author could have heard about the work of the other; or more interestingly we may be able to establish that the two authors were working along different lines. This is often clear from published material; but while papers are being written up there is time for second thoughts, and an author is not compelled to say with what end in view he began his investigations. Thus it appears that Lavoisier's studies of combustion arose out of an interest in effervescence, and that he first came across Priestley's work when he read his paper on making soda-water; and that Thomas

[1] H. Hartley, *Humphry Davy* (London 1966), 148; J. E. Jorpes, *Jac. Berzelius*, tr. B. Steele (Stockholm 1966), 59–60.

[2] On priorities, see N. Reingold, *Science in 19th-century America* (London 1966), 236ff. H. Guerlac, *Lavoisier—the crucial year* (Ithaca, New York 1961), 55; T. Wright, *Second Thoughts*, ed. M. A. Hoskin, (London 1968), 7–14.

Wright stumbled upon the theory that the stars in our galaxy are arranged roughly in the form of a grindstone in the course of his speculations on the position of Heaven and Hell. In these cases studies of the manuscripts—though Wright's chaotic papers could hardly be called laboratory notebooks—have made the printed sources more intelligible.

Laboratory notebooks can also help us to decide whether an experiment was really performed, and if so who did it. Particularly in the early part of our period there was little distinction between what would now be called a philosopher's example and a genuine experiment; Galileo had set an example of the use of 'thought experiments' in argument for the Copernican system and for the new dynamics which was no doubt followed.[1] In the latter part of the period authors did as a rule make it clear whether they were philosophising or describing actual experiments; here the problem is of 'cooking' or 'trimming'. These two processes, concocting observations to give the expected answer or merely adjusting genuine observations until they do, are familiar to every schoolboy, and have been not unknown in higher circles. These terms carry opprobrium, but trimming at any rate could be done in good faith before the rise of statistics in the middle of the nineteenth century. Until then there was no clear way of averaging a series of observations, and if one seemed very different from the others it could either be trimmed or omitted from the calculation; though in honesty it ought to have been published. The principle, too, that a chain is as strong as its weakest link had not been fully grasped, and we find series of experiments involving both weighing, which could be done very accurately, and volumetric analysis which could not, in which the result is given to a number of decimal places that would be appropriate to weighing alone. This specious accuracy can be

[1] A. Koyré, *Metaphysics and Measurement* (London 1968), essay III; on cooking, see C. Babbage, *Reflections on the Decline of Science in England* (London 1830), 174ff. For specious accuracy, see e.g. A. L. Lavoisier, *Elements of Chemistry*, tr. R. Kerr (Edinburgh 1790), 332ff, 338ff. H. Woolf, *The Transits of Venus* (Princeton 1959), 190.

investigated if we have the laboratory notebooks; as can the more deliberate cooking or trimming if the cook has neglected to destroy the evidence.

A notorious example of these processes, which need not involve us in moral judgements, is the controversy of the 1820s between Thomas Thomson of Glasgow and Berzelius over atomic weights.[1] Thomson was a convert to the view of William Prout that all atomic weights were whole multiples of that of hydrogen, while Berzelius rejected the corpuscularian view that all substances were composed of particles of the same matter in different arrangements, and thus was suspicious of Prout's theory. The question would seem to be a simple one; it involves making a series of analyses in order to see in what ratios by weight the chemical elements do combine. Thomson published numerous experiments showing that the ratios were simple, and that atomic weights were therefore integer multiples of that of hydrogen; while Berzelius' analyses indicated that they were not. After considerable polemic, Berzelius' values were confirmed by a third party; and for the most part his are closer to the modern figures. There were genuine difficulties about averaging, and it seems that Thomson accepted as the 'best' analysis those which led to the results he wanted, and as the 'best' analytical techniques those which gave them most readily. Berzelius, on the other hand, used generally accepted methods with great skill and arranged his less-discordant results without preconceptions.

Thomson's laboratory notebooks might cast light on his methods of calculation; and they might also indicate who did the experiments. For Thomson was a pioneer in the teaching of practical chemistry who, it appears, got his students to perform analyses for him to publish; conversely, earlier professors usually wrote theses for their students.[2] Later in the nineteenth century

[1] The published documents are reprinted in my *Classical Scientific Papers, Chemistry; 2nd series* (London 1970); and *Alembic Club Reprints, 20* (Edinburgh 1932).

[2] J. B. Morrell, 'The chemist breeders: the research schools of Liebig and Thomas Thomson', *Ambix, 19* (1972), 1–46; F. Stafleu, *Linnaeus and the*

the practice grew of employing assistants; busy men like William Crookes and Norman Lockyer seem to have thus done much of their scientific research by deputy. Only when their subordinates slipped up, as when Crookes' assistant supplied Lockyer's with a specimen that had not yet been purified, did the principal's reliance upon them become embarrassingly clear. There is no need for us to moralise about such practices, which no doubt in various forms continue. But since references to students or assistants in the introductions to books or papers are often most perfunctory when they should be most fulsome, it is only from anecdotes, published or in letters or memoirs, that we can get some idea of the prevalence of the use of assistants, and only from laboratory notebooks that we can find how much work they actually did. Account books can also help us here by indicating how much they were paid.

Laboratory notebooks may be filled in daily as work proceeds, or consist of loose sheets written up each day and subsequently bound together. But several books may be in use concurrently; and a man returning to a problem after an interval of weeks or years may return to the book in which his previous efforts had been written up.[1] A notebook may therefore show the gradual development of a man's ideas and experimental techniques for testing them; and it may also show him pursuing a line of investigation which he abandoned for reasons which he may make clear explicitly or implicitly. These things are often harder to find in published materials; for books and papers are usually ordered in a neat and logical way that makes the genesis of the idea hard to spot, and papers containing negative results are infrequently published. If we have the notebooks we may be able to confirm the suggestion that has been made of d'Alembert and

Linneans (Utrecht 1971), 143–4; W. H. Brock, 'Lockyer and the Chemists; the first dissociation hypothesis', *Ambix*, *16* (1969), 81–99, and 'Liebig's laboratory accounts', *Ambix*, *19* (1972), 47–58.

[1] *Faraday's Diary*, ed. T. Martin, 8 vols. (1932–36), preface to vol. 1; T. L. Hankins, *Jean d'Alembert* (Oxford 1970), 31; H. Hartley, *Humphry Davy* (1966), 60.

Davy, for example, that being ambitious and competitive they published papers too quickly, in a half-baked condition, in order to forestall others in the field; this would be shown to be less probable were we to find that they had spent months or years in thought and experiment on the topics of the papers in question.

To the laboratory notebooks of the physicist, chemist, or physiologist may be compared the notebooks of the natural historian; and natural history, it should be remembered, was in terms of the interest it aroused, the financial support it received, and the number of people engaged in it, among the most important of the sciences throughout our period. The *Notebooks* of Charles Darwin, which have recently been published,[1] show more clearly how his ideas were formed and how he set about collecting and assessing the evidence for evolution by natural selection. Since an author—especially if like Darwin he is collecting materials over twenty years—does not and cannot cite every author he reads and every piece of evidence he comes across, notebooks can add to what we find in the printed sources; particularly since authors on occasion do not mention in print those works which have been of most use to them. Natural historians also made journals after the manner of Gilbert White, and records of their travels; examples here being Linnaeus, Thomas Pennant and Joseph Banks. On their journeys natural historians made contacts with their fellows and exchanged information and specimens; their journals therefore show us contacts between men of science which may not be apparent in published works. Journals and notebooks may also help us to identify materials or plants inadequately or oddly described in printed works.

[1] In the *Bulletin of the British Museum (Natural History), Historical Series, 2* (1959–63) and *3* (1962–69), ed. G. de Beer and others. *Gilbert White's Journals*, ed. W. Johnson (London 1931); *Linnaeus' Tour in Lapland*, ed. J. E. Smith, 2 vols. (London 1811); *Pennant's Tour on the Continent, 1765*, ed. G. de Beer (London 1948); *Joseph Banks in Newfoundland and Labrador*, ed. A. M. Lysaght (London 1971); C. Lyell, *Journals on the Species Question*, ed. L. G. Wilson (New Haven 1970).

It is hard to separate these notebooks from commonplace books, or notebooks in which general reflexions, references to and quotations from books and articles, and odd memoranda are recorded. These again will be of most value for indicating what a man read, or perhaps who he talked with; and may disclose the first glimmerings of an idea which was later of great importance to him in the laboratory. Scientists' notebooks of this kind often contain much material which seems very little related to their research; Davy's notebooks at the Royal Institution, for example, are an extraordinary mixture and call out for an editor of very wide reading as well as the ability to read his scrawl. Similarly, in the commonplace books of those whom we do not think of as scientists there may be quite a lot of science. Thus Coleridge's *Notebooks*[1] include notes on chemical lectures by Davy, and also other matters connected with science; and Robert Southey also noted down curiosities of nature in much the same way as early fellows of the Royal Society had done. The presence of notes on science may help to cast light on a man's reading and hence on his poetry or political philosophy; to the historian of science it may help to indicate how familiar some scientific concept or discovery was, helping him thus to establish a norm. It will also remind him that the 'scientific community' has never been sharply delimited.

From notebooks and commonplace books the transition is easy to those notes made to clear the head, or to outlines or drafts of works which may or may not have been published. Thus in addition to Darwin's notebooks on transmutation of species, we have two drafts, of 1837 and 1842, setting out arguments for evolution which we find, differently arranged and much more elaborately argued for and supported, in the *Origin of Species* of 1859. These drafts, when compared with the *Origin*, indicate how Darwin's estimate of the relative importance of different lines of argument changed, and where further work, by him or by others, converted a plausible conjecture into a soundly based

[1] *The Notebooks of Samuel Taylor Coleridge* (New York 1957–); R. Southey and S. T. Coleridge, *Omniana*, ed. R. Gittings (Fontwell 1969); J. L. Lowes, *The Road to Xanadu* (Boston, Mass. 1927).

generalisation.[1] Similarly a manuscript by Faraday on force and matter—composed, curiously enough, soon after rather than before he had delivered an impromptu public lecture on this topic, which he had published—has recently been found; this must have been written in an effort to get clear his own thoughts, as he turned from billiard-ball atoms towards mere centres of force. Neither Darwin's drafts nor Faraday's brief paper were intended for publication; we may come across similar works intended only for their author, or perhaps for the perusal of a few friends also, which may illuminate later—or as in Faraday's case, earlier as well as later—publications.

Manuscripts in many ways similar to these are those which were intended for publication but which never got published. Newton, for example, wrote various drafts of prefaces, definitions and axioms for his *Principia Mathematica*, some of which he suppressed at a late stage in the preparation of the various editions of the book; these show how he refined the various concepts he used and give scope for conjecture as to why he dropped some passages to which he had devoted some time and trouble.[2] A man's annotations of his books can be similarly useful; if we can find an author's copy of a book in which we are interested we may find valuable notes in it. This is especially likely where there was not a subsequent edition of the book; for we may then find the author's second thoughts on the subject of which he has treated. Before our period, Harvey's initialled notes in his books are well known; and Davy, for example, made notes for a second edition of his *Elements of Chemistry* which never in the event appeared. People annotate not only books they have written, but others which they have owned or perhaps borrowed; Coleridge is particularly famous for this habit, so valuable for the historian

[1] C. Darwin and A. R. Wallace, *Evolution by Natural Selection*, ed. G. de Beer (Cambridge 1958); T. H. Levere, 'Faraday, Matter, and Natural Theology', *British Journal for the History of Science*, 4 (1968), 95–107.

[2] A. R. and M. B. Hall, *Unpublished Scientific Papers of Isaac Newton* (Cambridge 1962); and the variorum edition of the *Principia*, ed. I. B. Cohen and A. Koyré, 3 vols. (Cambridge 1971–3).

because it proves that the book was actually read even if the comments are not themselves very interesting. Even that minimum of annotation, putting one's name or a bookplate at the front of book, tells us something; we know for example that Dalton had at one time two copies of Newton's *Principia* because we have a note in a copy at Durham stating that it was his spare copy, which he therefore disposed of. Similarly if the book is a presentation copy this may be of interest; authors don't always indicate who the recipient is, but if they do this is evidence of respect or of close contact.

Drafts of unpublished material, and annotations to published work, lead us naturally towards drafts of books and papers which did get into print, and to which we turn next; while annotations of other people's writings point towards letters, which we must consider later. Learned societies often have the manuscripts of papers submitted to them which were subsequently published. In these days learned works are often published very slowly, but this is not, and has not been, always the case; Röntgen's paper on X-rays, for example, was submitted on 28 December 1895, and on 1 January 1896 he was able to send offprints to his colleagues.[1] Where publication is rapid, there is little time for second thoughts; while there may be misprints in the published version there will probably not be changes of substance from the manuscript that was read to the society or accepted by the editor of the journal. The majority of societies, particularly the most prestigious ones, have however rarely been very speedy in publication. The scientist wanting rapid circulation of his views or results sent them to *Rozier's Journal* in France, and to *Nicholson's Journal* or *The Philosophical Magazine* or, later, *Nature* or *Chemical News* in England. But the Royal Society in the nineteenth century

[1] W. R. Nitske, *Wilhelm Conrad Röntgen* (Tucson 1971), 6; *The Selected Correspondence of Michael Faraday*, ed. L. P. Williams, 2 vols. (Cambridge 1971), I, 480–1, 487; J. Davy, *Memoirs of the Life of Sir Humphry Davy*, 2 vols. (London 1836), II, 136; J. J. Waterston, *Collected Scientific Papers*, ed. J. B. S. Haldane (London 1928); J. Herapath, 'On the physical constitution of the Universe', *Annals of Philosophy*, 17 (1821), 274ff.

considered papers submitted to it as its property, even if it decided not to publish them; and various works subsequently considered very important thus disappeared into limbo. Why authors did not keep copies and submit them elsewhere is not clear; some did, notably John Herapath in his pioneer work on the kinetic theory of gases, but perhaps others were unwilling to antagonise the Royal Society.

When there was a delay in publication, the author would have had as a rule the chance to amend his manuscript or proof; and when we come upon a paper published in 1810 which purports to be the text of a lecture delivered in 1808 or 1809, we should handle it with due caution. If it appears to be remarkably prescient, this may be because it has been extensively revised. An extreme case of this delay was a lecture delivered in 1867 to the Chemical Society of London by Sir Benjamin Brodie, which was printed as a little book in 1880; here we have also a version taken down by a shorthand-writer and published in *Chemical News* eight days after it had been delivered.[1] Where the two texts differ, one would expect that the contemporary report is likely to be closer to what the audience actually heard. We know about this because both publications are correctly dated; but societies often fell behind in publication so that a journal dated 1750, for example, may not have appeared until 1751 or 1752 and what appears to have been rapid publication was in fact very slow. The Academy of Sciences in Paris was notorious for this in the eighteenth century; and papers by Lavoisier were extensively revised during the long gap between their acceptance and publication, but nevertheless came out bearing only the date of the year in which they were submitted. Comparison of the original manuscript with the published version can illuminate the process of revision. Some changes may have been made at the suggestion of the editor or of referees; again the manuscript may give us evidence of this.

Papers submitted to editors were, like official letters, often not

[1] On Brodie's lecture, see W. H. Brock (ed.), *The Atomic Debates* (Leicester 1967); C. C. Gillispie, *The Edge of Objectivity* (Princeton 1960), 229; H. Guerlac, *Lavoisier—the crucial year* (Ithaca, New York 1961), 61.

holograph manuscripts but fair copies made by an assistant or amanuensis. They will not therefore show, as a rule, the author's changes from his early draft; and if the paper is a joint production it will not be clear whether the authors have written different sections or whether, as with Ramsay and Travers' papers on the inert gases, when one put down the pen, even in the middle of a sentence, the other took it up and carried on.[1] Darwin's son described how his father wrote up his work and sent it to the local schoolmaster to be copied out neatly; he then corrected this copy, and the schoolmaster did a final fair copy for the press. No doubt many authors only made their amanuensis do one copy; but if we want evidence of how the work was composed we must look for the holograph manuscripts rather than the fair copies which may be all that the publisher or learned society will have. But it is easier when checking for changes made at the proof stage to have a handsome and clean manuscript with which to compare the printed version. We may hope that if a published work grew slowly through a number of drafts, the author will have preserved some of the earlier versions so that we can follow his mind at work; such drafts may, with his other papers, still be in the possession of his descendants, or have passed to a society with which he was associated, to his college or university, or to a library or museum. Here the rule of studying carefully what was published, before beginning on what was not, should not be broken.

A further problem about dating arises when discoveries are not written up for some time after they have been made, or when papers circulate in manuscript before publication. There may be various reasons for this; thus William Hyde Wollaston only published just before his death his process for making platinum malleable, though he had discovered it twenty-five years before. Competitors could only produce platinum in the form of a grey powder, so Wollaston had a lucrative monopoly in this useful metal—it was not in his day used in jewellery—and did not choose

[1] M. W. Travers, *A life of Sir William Ramsey* (London 1956), 192. F. Darwin, *The Life and Letters of Charles Darwin*, 3 vols. (London 1887), I, 153.

to publish his secret to the world.[1] Newton did not publish some of his discoveries for many years, but manuscripts did circulate among his friends; when Leibniz published his work on the differential calculus, of which Newton had grasped the principle many years before, he and his friends accused Leibniz of plagiarising from these manuscripts. The date of publication of Newton's works generally bears little relation to their date of composition. This raises a further problem for the historian, since those who were in Newton's circle would hear from him his latest views, and perhaps see them in manuscript; while those outside this group would have to be content with published sources in which Newton's view of matter, gravitation and aether, for example, might be very different from what he was currently expounding in private. In Newton's writings we have what might be called a private and a public chronology; and we also have this stage between composition and publication where papers circulated in manuscript.

Newton's case must be unusual, but at the end of the eighteenth century we find that two of the most prominent men of science in England, Henry Cavendish and Sir Joseph Banks, published little of their work.[2] Banks, like Newton, allowed his manuscripts to circulate; and seems not to have worried overmuch if others

[1] W. H. Wollaston, 'On making platina malleable', *Philosophical Transactions, 119* (1829), 1–8; we need a study of him and his family. On Newtonian chronology, see A. Thackray, *Atoms and Powers* (Cambridge, Mass. 1970), esp. ch. II.

[2] On Banks, see note 1, p. 77 and J. C. Beaglehole (ed.), *The Endeavour Journal of Joseph Banks*, 2 vols. (Sydney 1962); H. Cavendish, *Electrical Researches*, ed. J. C. Maxwell (Cambridge 1879); for sketches of Banks and Cavendish, see J. Barrow, *The Royal Society* (London 1849). Banks as President of the Royal Society was in the tradition of Hans Sloane; see *The Sloane Herbarium*, ed. J. E. Dandy (London 1958), and *Index to the Sloane MSS*, ed. E. J. L. Scott, 2nd ed. (London 1971); F. C. Sawyer, 'A short history of the libraries and list of MSS and original drawings in the British Museum (Natural History)', *Bulletin of the B.M. (Nat. Hist.), Historical Series, 4* (1971), 77–204; P. Mander-Jones, *MSS in the British Isles relating to Australia, New Zealand, and the Pacific* (Sydney 1973). N. Matthews and M. D. Wainwright, *Guide to MSS and documents in the British Isles relating to Africa*, ed. J. D. Pearson (London 1971).

incorporated matter from them into their own books and papers. It is only in our own day that Banks' splendid journals and collections, which found their way into the British Museum (Natural History), have begun to be published; and this is leading to a re-establishment of his reputation as a natural historian; the view that he was little more than a domineering aristocrat is quite untenable. Banks seems to have been uninterested in publishing even work that would have required very little if any polishing to prepare it for the press; and the same was true of Cavendish, many of whose electrical researches remained unknown until they were published, edited by the great physicist Clerk Maxwell, seventy years after his death. By then, naturally, the original discoveries had been rediscovered by others; these papers had not been seen by Cavendish's contemporaries, and the experiments do not seem to have been known to them either. To make discoveries, write them up, and not communicate them to anybody is unusual, and Cavendish was an unusual man; but there must be many scientists who have published less than their total *opus*, and whose work can only be adequately comprehended if the manuscripts as well as the published material are studied. The historian must then try to decide how much of the manuscript material was known to contemporaries; and here to know that Banks kept open house and was a clubbable man while Cavendish was a recluse is helpful. Whether what a man knows but keeps to himself can be described as science, is a question different people would answer differently; but as a rule only public knowledge would qualify. Manuscript material may help us to resolve apparent inconsistencies in a man's writings by bringing out how and why he changed his mind; and we should remember that Davy, Alexander von Humboldt, and Goethe are among those who have inveighed against consistency, as a sign of intellectual ossification.

A man's opinions of other people's work may be found in his published writings, especially reviews, or perhaps more pungently in his annotations of books or papers if he had the useful habit of annotating. But it is in his letters that we may hope to see his views most informally expressed, and in correspondence we do sometimes see some new concept being worked out or some new experiment being planned. The historian is fortunate when the principals in some new movement live in different places, so that instead of talking things over, and reminiscing on paper only long afterwards, they have to exchange their thoughts by correspondence. There is an immediacy about the letters of Faraday and William Whewell discussing nomenclature that is missing from Faraday's paper in which he proposed the new names which his researches had made necessary;[1] and similarly about the letters between Archdeacon Thorpe in Durham and Bishop van Mildert in London about the setting up of Durham University.

Letters can also be a valuable source of gossip, but it should not be taken for granted that because something is in manuscript it must be true. Thus there is a letter, recently published,[2] from Gerrit Moll of Utrecht to Faraday, alleging that in the Napoleonic period Gay-Lussac and Thenard, rivals of Davy, had prevented the publication of his papers in the *Annales de Chimie* and had threatened the editor of the *Annales de Physique* that his journal would be closed down by the police if he published Davy's work, which he nevertheless did. The story had some credibility because through their patrons Laplace and Berthollet, Gay-Lussac and Thenard had the ear of government; but examination of published sources shows it to be baseless, since Davy's papers did appear in the *Annales de Chimie*. La Metherie, editor of the *Journal de Physique* became embittered, not to say paranoid, in his old age; and Dutch scientists had little cause to love the Napoleonic regime; that a story like this could circulate is as interesting to the historian as if it were true.

[1] L. P. Williams (ed.), *Selected Correspondence of Faraday* (Cambridge 1971), I, 265–72; Thorpe's papers are preserved in Durham University Library, and see *Durham University Journal, 25–27* (1926–32), for extracts from this correspondence; C. E. Whiting, *The University of Durham, 1832–1932* (London 1932).

[2] M. P. Crosland, 'Humphry Davy', *British Journal for the History of Science*, *6*, (1973), 304–10.

What we may call the private correspondence of an important scientist may thus tell us much about his own development, his friends, the gossip of the day, the working of scientific societies and of pressure-groups within them, and the context of science generally; especially if we have got both sides of the correspondence. But some men of science have had, in addition to this, correspondence of what we may call a more public nature. Henry Oldenburg, the first secretary of the Royal Society and editor of its journal, maintained an extensive foreign correspondence and thus kept the Society in touch with groups and individuals advancing science abroad.[1] As societies developed, they took over these functions increasingly through their journals; but speedier and more informal communication by letter went on, as it still does, between those in different countries or different parts of the same country. And as Oldenburg had done, an important man would find himself acting as a focus for the collection and distribution of information. This could easily happen particularly in natural history, where a man such as Linnaeus or Sir Joseph Banks had access to splendid collections including a number of type specimens from which the species had been named and with which a doubtful specimen could be compared. Botanists enriched their collections by exchanging spare specimens; and the correspondence of important natural historians is a rich field. Since such men were in our period as a rule attached to some institution of higher learning, scientific society, or botanical garden, their correspondence will belong in part to their biography and in part to the history of the institution which they served or manipulated.

Having a similar official character is the correspondence of those pundits to whom government turned for advice on

[1] A. R. and M. B. Hall (ed.), *The Correspondence of Henry Oldenburg* (Madison, Wis. 1965–); W. R. Dawson, *The Banks Letters* (London 1958); E. Forbes (ed. and tr.), *The Euler-Mayer Correspondence, 1751–55* (London 1971); E. C. Herber (ed.), *Correspondence between S. F. Baird and L. Agassiz* (Washington, D.C. 1963); M. B. Hall, 'Oldenburg and the Art of Scientific Communication', *British Journal for the History of Science, 2* (1965), 277–90.

scientific questions. When, as in France and in Prussia, there was an Academy of Sciences part of whose duty it was to advise government, requests for advice went to them and were considered by a committee or an individual member. In Britain things were less formal; the President of the Royal Society throughout our period might find himself being consulted on a wide range of questions, which he might answer himself or pass on to somebody with expert knowledge. Banks' correspondence contains numerous letters relating to matters of this kind, especially connected with agriculture and with voyages of discovery; and a few days after Banks' death in 1820 Sir Humphry Davy, though he had not yet been elected President of the Royal Society and all his work had been in chemical science, was consulted about the merits of a candidate for the post of Superintendent of the Ordnance Survey.[1] Davy had a few years earlier been approached, as a pundit, by a group of coal-owners anxious to prevent explosions in their mines, and had invented the safety-lamp; he had also advised the government, less happily, on the ventilation of Newgate Prison and of the House of Lords. As President of the Royal Society, he received numerous official requests for advice; most notably on how to protect copper-bottomed ships from corrosion, but also on a whole range of things vaguely connected with science. Such official correspondence of Presidents of the Royal Society may usually be found there, but has often got among the man's other papers; and we may have little record of the informal verbal advice for which there was considerable opportunity in an age when public men were likely to belong to the Royal Society, and men like Robert Peel to be elected to its Council.

The President of the Royal Society occupied a prominent but not an official position; in 1818, on the reconstituting of the Board of Longitude, an attempt was made, in Sir John Barrow's phrase, 'to open salaried office to men of science' but 'in a subsequent fit of economy, this poor pittance was withdrawn'.[2]

[1] C. Close, *The Early Years of the Ordnance Survey*, ed. J. B. Harley (Newton Abbot 1969), 84.

[2] J. Barrow, *The Royal Society* (London 1849), 62; for a different view, see

Newton, and in the nineteenth century Sir John Herschel and Thomas Graham, held the position of Master of the Mint; but through most of our period the Astronomer Royal acted as adviser to government on a whole range of topics, some of them very loosely connected with his primary duties at Greenwich Observatory. The correspondence of such a man as George Biddell Airy, Astronomer Royal through most of Queen Victoria's reign, can tell us a great deal about the kind of questions that were being discussed and about interdepartmental struggles between, for example, the Department of Science and Arts at South Kensington and the Royal Observatory in the competition for funds and power. Government grants for scientific research in universities in England only began in the last two decades of the nineteenth century, though for some time before that grants had been made to the Royal Society to be distributed to those in need of them; for most of our period government grants came particularly for expeditions, and naturally the Astronomer Royal and the President of the Royal Society were consulted about suitable people to go as astronomers or natural historians. Funds were also required to equip those chosen with apparatus, and to see their reports through the press when they got home; and for these reports the cooperation of experts in the British Museum or elsewhere was required. The papers of the Astronomer Royal are one place where one can investigate what went on; once again after getting all that one can from the printed materials.

We mentioned Davy's being consulted even before his election as President of the Royal Society; and in the nineteenth century there were various pundits who although they held no official position were nevertheless often turned to formally or informally for advice by those in power. Thus during the Crimean War, Faraday was consulted about a proposal to smoke the Russians

---

C. Babbage, *Decline of Science in England* (London 1830), 66–100. For instructions for a scientific voyage, see W. H. B. Webster, *A Voyage . . . in H.M. Sloop Chanticleer*, 2 vols. (London 1843), II, 369ff.

out of Kronstadt with sulphur-filled fireships. But Sir John Herschel became the man to whom anybody in need of sound advice on any topic connected with science turned in the middle years of the nineteenth century. After his defeat by the Duke of Sussex in the contest for Presidency of the Royal Society in 1830, Herschel had gone to the Cape of Good Hope and compiled a catalogue of the southern stars and nebulae.[1] On his return he decided deliberately not to specialise; his wide reading and acquaintance made him a person of great importance, his reviews and general works were widely read and highly praised, and his papers are correspondingly more rewarding than those of men who stuck closer to their last.

A younger contemporary of Airy and Herschel was George Gabriel Stokes, who was for over thirty years Secretary of the Royal Society and editor of its *Philosophical Transactions*. He had made important studies of viscosity and of fluorescence; but he devoted the years of his maturity chiefly to his editing; advising authors how to improve papers they had submitted to the Royal Society and corresponding with referees. The correspondence of those holding positions like Stokes' form a transition between the private papers that cast light on the life and times of an individual, and the documents which relate to scientific or other institutions, and to government departments; though naturally no fixed line can be drawn between the two classes. Correspondence, manuscripts of papers printed and of addresses delivered, and laboratory notebooks have already been mentioned; and we can now begin to consider less personal documents.

The minute books of the committee or council of a society will often tell us tantalisingly little about what went on in their meetings. Written up by a tactful secretary and rarely indexed, minutes can conceal a violent altercation behind a bland recording of a decision. We may find that a meeting had to be adjourned, or that somebody withdrew from an election or from the

[1] D. S. Evans *et al.*, (ed.), *Herschel at the Cape; diaries and correspondence, 1834–8* (Austin, Texas 1969); L. P. Williams (ed.), *Selected Correspondence of Faraday* (Cambridge 1971), II, 749–51.

society; but we shall not as a rule find out why.[1] But other questions can be answered; we can find out how assiduous the various members of the committee were in their attendance, which will give us some clue as to how dedicated they were to the society, and tell us which were the people who really ran the society. In learned societies the President, Treasurer and Secretary with one or two others often make most of the decisions, which the committee or council approves and sometimes amends, until some crisis blows up. Minutes can also help to indicate the relative importance of the various sciences at a given time, by showing who got the most influential positions; that is, as a foundation for a prosopography. The minutes of the Royal Institution in London have recently been used to controvert the received view that the Royal Institution was set up, chiefly by Count Rumford, as a technical college for mechanics, and only later became a fashionable centre for lecturing and advanced research. Closer study of the documents, it is claimed, indicates that from the first the most important group of backers and office-holders were 'improving landlords', who were never very sympathetic to the mechanics, and got what they wanted in Davy's research and lectures on agriculture and tanning. Rumford, on this view, was important chiefly as a publicist and was regarded by at least some of his colleagues as a mountebank whose departure to France was welcome.

Because scientific societies published in their journals most of the scientific communications which they received, and because their meetings were also reported in more popular journals, we may not find many surprises on looking at their manuscripts; though we may find out things about the detailed workings of the society, and about its social composition and function, by a close study of manuscript materials. We are more likely to be surprised when we examine the archives of bodies not directly

[1] L. F. Gilbert, 'The Election to the Presidency of the Royal Society in 1820', *Notes and Records of the Royal Society*, *11* (1955), 256ff. M. Berman, 'The Early Years of the Royal Institution', *Science Studies*, *2* (1972), 205–40; A. H. Church, *The Royal Society Archives*, 2 vols. (London 1907–8), goes up to 1806.

scientific: universities and colleges, government departments, and companies engaged in trade or manufacture. We can from these documents learn something about who patronised science, and why; and about how science was taught and what the syllabus was. Throughout our period the extent to which science was taught or pursued in a university has been taken—particularly by scientists and their publicists—as a touchstone of activity and modernity. There is therefore much material, published and unpublished, concerning syllabuses and courses; we shall consider in later chapters the printed handbooks and reports of commissions, and here mention only manuscripts. Syllabuses, and lists of examiners and of successful candidates, were usually published; but everybody knows that what is taught may bear little relation to what the syllabus lays down—indeed the best syllabuses may be so loosely written as to give ample scope to the individual lecturer. A better indication will probably be found from examination papers, a neglected source for study of the diffusion of science. But best of all are lecture notes, either those compiled by the lecturer or taken down by his hearers; these may be found at the institution where they were delivered, but may be discovered almost anywhere.[1]

What should be found at the college or university are the minute books in which decisions about the science courses are recorded, and the account books which reveal how much was spent on apparatus and buildings and what salaries were paid. The laboratory account books of Justus von Liebig at Giessen, where he began laboratory instruction in chemistry, have been used to investigate the problem why he developed a world-famous research school while his contemporary Thomas Thomson at Glasgow did not; Liebig was able to screw money out of the government of Hesse-Darmstadt in a way that no British scientist

[1] T. Cochrane, *Notes from Dr. Black's Lectures on Chemistry, 1767/8*, ed. D. McKie (Wilmslow, Cheshire 1966); J. Walker, *Lectures on Geology*, ed. H. W. Scott (Chicago 1966); J. B. Morrell, 'Science and Scottish University Reform: Edinburgh in 1826', *British Journal for the History of Science*, 6 (1972) 39–56.

could, though the amounts which even he screwed were very small.[1] Account books, by indicating what was bought in the way of apparatus and equipment, may cast light on the policy of the institution buying it; we can see for instance whether they bought expensive and prestigious batteries, electro-magnets or telescopes, or whether they relied by choice or necessity on sealing-wax and string. We may be able to see how much was spent on buildings and how much on apparatus, books and salaries; we must then of course try to account for these ratios, remembering that our function is not to give summary judgement on a period but to study it critically.

Universities and government departments shade into one another; particularly in France and Germany, but also in Britain in the complex of institutions at South Kensington which became Imperial College, and in the Royal Commissions on the Universities of Glasgow, Oxford, Cambridge and Durham, and the various reports on the University of London. Published writings and reports on these various institutions will be mentioned in the appropriate chapter. The letters and papers of such men as T. H. Huxley and Norman Lockyer can tell us something about the functioning of the South Kensington institutions: Huxley archives being at Imperial College, and Lockyer ones at the Norman Lockyer Observatory at Sidmouth. In Britain, the government thus supported a school of mines, a college of chemistry, and a college for teachers; and the papers of these institutions, now Imperial College, can cast light on the relations between government and science.[2] They were accommodated on

[1] J. B. Morrell, 'The Chemist Breeders', and W. H. Brock, 'Liebig's Laboratory Accounts', *Ambix*, *19* (1972), 1–46, 47–58; on electromagnets, N. Reingold (ed.), *The Papers of Joseph Henry* (Washington, D.C. 1972–), I, 312ff, 373ff.

[2] J. Pingree, *List of the papers and correspondence of Lyon Playfair in the Imperial College Archives* (London 1967); *Guide to the Public Records*, 2 vols. (London 1963); J. Lawson and H. Silver, *A Social History of Education in England* (London 1973); J. Bentley, 'The Chemical Department of the Royal School of Mines' and 'Hoffman's return to Germany', *Ambix*, *17* (1970), 153–81, *19* (1972), 197–203; W. G. Stephens and G. W. Roderick, 'Private enterprise and chemical

the site at South Kensington bought with the proceeds of the Great Exhibition of 1851, where they were joined first by the British Museum (Natural History) and then by other museums of scientific importance. The British Museum had long been an important institution in the field of natural history, particularly when it acquired Joseph Banks' materials and his librarian, the great botanist Robert Brown, to look after them. Davy, who was President of the Royal Society in the 1820s, urged the government to make the British Museum into a research establishment like Cuvier's Museum at the Jardin des Plantes in Paris; but this did not happen until the mid-century with the separation of natural history from the museum's other activities. Because little research went on in the universities during most of the nineteenth century, the British Museum, the gardens and observatory at Kew, and private institutions like the Royal Institution are very important; and their collections and archives merit a great deal of study.

The government also supported expeditions; papers relating to these may be found at the Royal Greenwich Observatory at Hurstmonceaux, where the archives of the Board of Longitude are to be found. This body, which brought together scientists, naval men, and representatives of government, was dissolved when chronometers and lunar observations were deemed to have proved satisfactory means of determining longitude in the 1820s; but papers relating to later expeditions may also be found at the observatory. The logs of ships are as a rule in the Public Record Office, and there also may be found the manuscript journals made by officers, and occasionally seamen, on voyages of survey and discovery, which had to be handed to the captain before the ship docked. The Maritime Museum at Greenwich has plans of ships, and collections of official letters concerning the dispatch of expeditions; the hydrographic department of the

---

training in 19th century Liverpool' and 'Education and training for English Engineers', *Annals of Science*, 27 (1971), 83–94, 143–64; P. Sviedrys, 'Physical Science at Victorian Cambridge', *Historical Studies in the Physical Sciences*, 2 (1970), 127–51; W. D. Miles, 'W. J. Macneven and early laboratory instruction in the U.S.A.', *Ambix*, 17 (1970), 143ff.

Admiralty has charts; and at the British Museum are more charts, letters, and journals.[1] The Scott Polar Research Institute at Cambridge has documents concerned with polar exploration. Anybody contemplating research involving voyages should at the outset consult the various publications of the Hakluyt Society and the Navy Records Society, as examples of editing and for their bibliographies.

Surveys nearer home were also supported, most notably the Ordnance Survey; in its early years, from about 1790 to 1840, the survey had close connexions with the world of science, and the Geological Survey grew out of it. An excellent guide to the sources available for studying this body has been prepared by J. B. Harley to a recent reprint of a standard history;[2] and this would repay the attention of historians of science who have concentrated too much in the past on theoretical physics and laboratory experiments. In the United States there was a comparable body, the Coast Survey, which soon extended its field inland to the west and had become by the civil war the largest employer of scientists in the U.S.A. Previous historians of American science had stressed its provincial character, and remarked—perhaps in whiggish vein—how despite propaganda and money no American Helmholtz, Clerk Maxwell or Pasteur had appeared; however, by studying letters and documents, including those connected with the survey, Nathan Reingold has been able to sketch a picture of what American scientists actually did, which was mostly geography, natural history, and geology of a descriptive kind. This is of course not surprising in a new country being opened up; a similar emphasis, though less extreme, had characterised European science a little earlier and has always appealed to governments.

Those companies, like the Dutch and English East India Companies and the Hudson's Bay Company, which enjoyed territorial jurisdiction, or like the African Company which had trading

[1] A. Day, *The Admiralty Hydrographic Service, 1795–1919* (London 1967); G. S. Ritchie, *The Admiralty Chart* (London 1967).

[2] C. Close, *The Ordnance Survey*, ed. J. B. Harley (Newton Abbot 1969); N. Reingold, *Science in 19th-century America* (London 1966), 152ff.

interests in a little-known area, also—sometimes fitfully—supported exploration and the study of natural history.[1] The maps drawn up by Major Jones Rennell in Bengal were models of their kind and probably superior to anything done in Britain, and for them he was awarded the Copley Medal of the Royal Society in 1791. The Botanical Gardens at Calcutta, founded 1788, became famous particularly in the time of William Roxburgh who had splendid illustrations of the Indian flora made by local artists under his supervision; and later there were close connexions between the garden and that at Kew, when the Hookers, father and son, who directed Kew were particularly interested in Indian plants. The zoologists of Europe were startled when the American Thomas Horsfield, under the patronage of Stamford Raffles who became first President of the Zoological Society of London, described the Malay tapir which belongs to a genus which had been previously thought by Europeans to be confined to America. In the India Office Library are journals and drawings of natural history; which again have not been enough studied by historians of science. The Hudson's Bay Company encouraged some exploration, though like the Dutch East India Company it seems to have been prone to regard the results of such exploration as secrets to be carefully guarded; but the expeditions of Hearne and Mackenzie, of Governor Simpson and his nephew, and of John Rae, are noteworthy, and the company cooperated with

[1] W. Foster, *A Guide to the India Office Records, 1616–1858* (London 1919); M. Archer, *Natural History Drawings in the India Office Library* (London 1962); *Icones Roxburghianae* (Calcutta 1964–); F. Merk (ed.), *Fur Trade and Empire: George Simpson's Journal*, 2nd ed. (Cambridge, Mass. 1968); G. E. E. and E. H. J. Feeken, and O. H. K. Spate, *The Discovery and Exploration of Australia* (Melbourne 1970); on the African Company, see W. E. F. Ward's introduction to the reprint of T. E. Bowditch, *Mission from Cape Coast Castle to Ashantee* (London 1966); on exploration generally, R. A. Skelton, *Explorer's Maps* (London 1958), and R. V. Tooley, C. Bricker, and G. R. Crone, *A History of Cartography* (London 1969). On the African Association, E. W. Bovill, *The Niger Explored* (London 1968). On exploration and secrecy, M. Sauer, *Expedition to the Northern Parts of Russia* (London 1802), appendix V. On the East Indies, see p. 168, note 1.

the official or semi-official expeditions of John Franklin, John Richardson and George Back. Correspondence and journals about such expeditions repay study: as do those about the exploration of Australia and of Africa by a mixture of official, semi-official, private or missionary expeditions in which science was often not prominent.

Manufacturing companies have had their connexions with science; this applies especially to instrument makers and to companies involved with the high technology of their day, but also to chemical and metallurgical industries. There is room for the historian of science, as well as for the historian of technology, the industrial archaeologist and the economic historian in this field.[1] The correspondence of industrialists and the account-books of their companies are valuable sources, and so is the library of the Patent Office; and much can be learned from the architect's drawings for factories about lighting and sanitation, working conditions, stockholding and processes, as well as about the engineering needed for building the great mills. But since technological advances were frequently not published by being patented, patents being hard to defend, some of the most valuable documents are close to what would now be called industrial espionage. Diaries or reports, usually by expert foreigners, of their visits, open or clandestine, to factories, mines or railways are

[1] There is a useful series of books on industrial archaeology published by Longmans. See the *Transactions* of the Newcomen Society, and the journal *Industrial Archaeology*. J. P. Tann, *The Development of the Factory* (London 1970); B. Woodcroft, *Alphabetical Index of Patentees of Inventions* (1854, reprint London 1969). G. Head, *A Home Tour through the Manufacturing Districts* (London 1836); W. O. Henderson (ed.), *Industrial Britain under the Regency, 1814–18* (London 1968); C. von Oeynhausen and H. von Dechen, *Railways in England, 1826 and 1827*, tr. E. A. Forward, ed. C. E. Lee and K. R. Gilbert (Cambridge 1971); N. Rosenberg (ed.), *The American System of Manufactures* (Edinburgh 1969); E. T. Svedenstierna, *Tour of Great Britain, 1802–3; the travel diary of an industrial spy* (Newton Abbot 1973). A useful reprint is J. Farey, *A Treatise on the Steam Engine* (1827), 2 vols. (Newton Abbot 1971); the second volume was never published, and is reprinted from proofs at the National Reference Library of Science and Invention.

an important source of methodical and critical observations which tell us things that might not have seemed remarkable to those working there, or which they might have preferred to keep to themselves. Such reports also cast light on the relations, social and technological, between different countries. They may be found in the archives of the company or government which sent the observer, or among his own papers—for a scientist on tour abroad would usually try to see interesting local industries on his own account—or they may have been published in whole or in part, in a book, a journal, or an official report. It is to these printed sources that we must next turn; beginning with journals, which have played so prominent a role in the development of modern science.

CHAPTER 4

Journals

We have already noted that the scientific journal began with the setting up of the scientific societies of the 1660s, and that its appearance marked a new phase in the history of science. The historian wanting to study almost any aspect of science will find in journals a most important source; and except for the last part of our period—from about 1850—there were very few journals which were too recondite for the layman to enjoy. Thus we find that not active 'scientists' alone, but others interested in various degrees in science would have come across papers in scientific journals; we should not forget that,[1] like those elsewhere, 'the English Romantics were heirs of their country's scientific revolution quite as much as the English inventors, who turned it into an industrial revolution'. Coleridge and Southey were avid readers of scientific journals; and even at the end of our period the general scientific journal *Nature* had, as it still has, a readership by no means confined to men of science. Conversely, many publications which we might be disposed to consider as devoted to the arts, such as the *Gentleman's Magazine*, the *Ladies' Diary*, and the *Edinburgh*, *Westminster*, and *Quarterly Reviews*, carried important articles on the state of science. Here, as usual, we must be careful not to draw the boundary of 'science' so firmly as to exclude much of what would have seemed to be science to contemporaries.

We find this same difficulty when we look at secondary journals in the history of science; for papers may appear in a whole range

[1] S. T. Coleridge and R. Southey, *Omniana*, ed. R. Gittings (Fontwell 1969), 18; J. O. Hayden, *The Romantic Reviewers, 1802–24* (London 1969); see my review-article in *History of Science*, *9* (1970), 54–75; and W. D. Wetzels, 'Aspects of Natural Science in German Romanticism', *Studies in Romanticism*, *10* (1971), 44–59.

of specialised and general journals concerned with the history of science, or of particular sciences, and may also be found in publications chiefly concerned with science, social history, philosophy or literary criticism. Fortunately the journal *Isis* publishes an annual bibliography of the whole subject, which is a useful point of departure. A publication devoted explicitly to problems and sources is *History of Science*, while for those interested in sources for the study of natural history there is the *Historical Series* of the *Bulletin of the British Museum* (*Natural History*), which has published Darwin's notebooks and various descriptions of collections of specimens; and the *Journal* of the Society for the Bibliography of Natural History. This society has also published facsimiles of important but rare items. The *Dictionary of Scientific Biography* includes references to articles which were recent when the biographies were written.[1] Bibliographies in useful papers will give a guide to other sources, primary and secondary, in journals. Experience will soon show which journals are most likely to contain useful papers in a given field; some, like the triennial, multilingual *Proceedings* of the International Congresses on the History of Science, must be of use to nobody though one day they will be valuable to historians of the history of science. The value of articles in secondary journals is twofold; on the one hand they indicate the whereabouts of material, and the state of research, on the topic in question; and on the other they suggest how to handle it. The same article will often fail to perform both functions; and it is valuable to read papers in fields removed from one's own in order to see what use can be made of sources. Thus the historian chiefly concerned with nineteenth-century chemistry may well find most suggestive a paper on eighteenth-century physics or on nineteenth-century biology. We tend to resist novel interpretations in those fields which we know most about; it is galling, in science or in history, to have to revise a synthesis painfully arrived at in order to incorporate new insights, but the process is made easier if one can

[1] C. C. Gillispie (ed.), *Dictionary of Scientific Biography* (New York 1970–); also T. I. Williams (ed.), *Biographical Dictionary of Scientists* (London 1969).

carry the new ideas across from another field, rather than meet them head-on.

After the reading of some secondary papers, to which references will have been found in bibliographies of one kind or another and which will themselves lead back to further papers, the time will come to plunge into the primary journals. When this time comes it is difficult to say; on the one hand, time spent in reconnaissance is seldom wasted, but on the other life is short, and secondary publications are multitudinous and often repetitive. The right moment has been picked when the sources seem fresh and can be read without blank incomprehension, though affording the sting of surprise. Thereafter the historian must read both primary and secondary sources; the temptation to confine himself to original documents must be resisted, (though the golden mean lies on that side), for this leads to antiquarianism rather than to history.

Where one begins in the primary journals will depend upon the period in which one is interested, and to a lesser extent upon the branch of science with which one is most concerned. For it is best to begin with general journals, covering the whole field of science; from them one can get some idea of how the sciences seemed to be related to each other, and perhaps of what it felt like to be a man of science of the period. The temptation is to pursue the publications of some one person, probably following up footnotes in secondary sources or using the very valuable, though never-completed, *Royal Society Catalogue of Scientific Papers,*[1] which is a splendid index, by author and in some fields by subject, to the major scientific journals up to the end of the nineteenth century. But if we just follow up some favourite author—favourite perhaps only because nobody has yet done a thesis on him—we cannot see him in his proper context. Reconnaissance among the secondary sources is valuable; among the primary sources it is a necessity; and we must not grudge the time

[1] Royal Society of London, *Catalogue of Scientific Papers, 1800–1900*, 19 vols. (London 1867–1925); then see J. D. Stewart *et. al.*, *British Union Catalogue of Periodicals*, 4 vols. and supplement (London 1955–62).

spent in thus browsing, or set ourselves some limited project before we have seen how the sources bear upon it. Blinkers are only of value when the road ahead is clear and distinct.

Those entering the history of science are all too prone to attribute to their particular hero all sorts of discoveries, some of which he usually did not make, and others of which would not have seemed very interesting to him or his contemporaries; and to miss the discoveries or interpretations which did seem striking at the time because they seem remote from our current preoccupations. Thus Sir Harold Hartley's excellent brief life of Sir Humphry Davy,[1] written by a man with great knowledge of the development of electrochemistry, is scrupulously fair in indicating where Davy made mistakes or erroneous interpretations of his data. But it is written with benefit of hindsight, and these 'errors' are not always related as they might have been to the received notions of the day. Again, Hartley could make little of Davy's last book, *Consolations in Travel*; the preoccupation in that work with materialism means that it must be approached through eighteenth-century authors like David Hartley and Joseph Priestley, and through Romantic authors such as S. T. Coleridge. A *Life* of Davy in which a great deal is not left out—not detail, that is, but important facets of his character—has yet to be written. The author will have to be interested in social climbing in the early nineteenth century as well as in electrochemistry, natural philosophy, natural theology and some Romantic writings. To do this as well as Hartley executed his limited task will be very difficult.

To see the state of science at a period—how it was organised, what fields were most active, and where the frontiers were drawn—we therefore look first at the general journals. This is easiest in the first half of our period; for at first there were only general journals, associated with the scientific societies and academies. The *Philosophical Transactions* of the Royal Society began the tradition of publishing signed papers in the vernacular and had a

[1] H. Hartley, *Humphry Davy* (London 1966); for a different view of Davy, see T. H. Levere, *Affinity and Matter* (Oxford 1971).

procedure for refereeing the papers. The *Transactions* began as the enterprise of the secretary, Henry Oldenburg; for a time their publication was interrupted, and Robert Hooke edited *Philosophical Collections* instead. Only after the 47th volume had appeared did the Royal Society in 1752 make itself responsible for the *Philosophical Transactions*: and even then an 'Advertisement' appeared at the beginning of each number warning readers that papers did not in any sense represent the opinions of the Society as a body. The early volumes contained book-reviews and brief letters from foreign correspondents as well as scientific papers; but they did not include regular reports of the meetings of other societies or academies.

The papers themselves make fascinating reading, leading into all sorts of by-ways. Sir Robert Moray, the first president, distinguished himself with a paper repeating the old story that geese grew out of barnacles; but the general impression is not of poor observation but of lack of system. This was no doubt a characteristic of the science of the day; Boyle, Huygens and Newton were exceptional men, and we must not judge the norm from their productions. Weather-reports, analyses of mineral-waters, and accounts of freakish or unusual happenings fill many of the pages; there are descriptions of alarming thunderstorms, of exotic creatures, and of monstrous births and unusually large bladderstones. The science is amateurish and often uncritical, not to say desultory; the papers are often compulsive reading, but sometimes pedantic. These aspects of the science of the early Royal Society were satirised by Swift in *Gulliver's Voyage to Laputa*, and denounced by the dramatist and populariser 'Sir' John Hill in 1751.[1] Perusal of the *Philosophical Transactions* makes it clear why the men of science who had been feared as possible innovators in politics and religion in the 1660s had by the turn of the century come to be seen for the most part as learned dunces: the absent-minded professor had appeared on the scene.

Most of the early papers, then, made few demands upon the

[1] See J. Hill, *A Review of the Works of the Royal Society* (London 1751); M. Nicolson, *Science and Imagination* (Ithaca, New York 1956), essay V.

reader, except sometimes for the willing suspension of disbelief. But papers on astronomy, optics, and some aspects of biology must have required much more effort, and are evidence of the developed state of those sciences, showing both theoretical and observational or experimental progress, with contributions from different countries. In the course of the eighteenth century, the papers in the *Philosophical Transactions* became increasingly austere and rigorous; by the early nineteenth century, the journal had acquired a large quarto format, most papers were much longer than they had been in the early days, and there is a feeling of dignity about the publication even when it includes articles on fairy rings or why the eyes in portraits follow one around.[1] Both these papers were from the pen of an established chemist and metallurgist of a notable scientific family, William Hyde Wollaston; and by the nineteenth century it was difficult to get a paper of a light or speculative kind into the *Philosophical Transactions* unless one had a reputation already made. For the nineteenth century, therefore, this journal no longer represents a norm; it was a prestigious vehicle in which men already of some note published long and often important papers, and which even rejected work submitted by the elderly Dalton. In the first half of the century, most papers in this journal would still have been intelligible to the non-experts, who indeed composed the majority of the Royal Society; but the dilettante would frequently have found in these weighty tomes more than he wanted to know.

The role of a popular general journal which the *Philosophical Transactions* had filled in the early days was from the end of the eighteenth century played by various private publications. Thus in Britain we find the *Philosophical Magazine*, founded by Alexander Tilloch in 1798, and William Nicholson's *Journal of Natural*

[1] W. H. Wollaston's papers come in *Philosophical Transactions*, *97* (1807), 133–8, *114* (1824), 247–56; E. H. Gombrich, *Art and Illusion* (London 1960), 234, 354; D. C. Goodman, 'Wollaston and the Atomic Theory of Dalton', *Historical Studies in the Physical Sciences*, *1* (1969), 37–59; A. L. Smyth, *John Dalton, 1766–1844: a bibliography* (Manchester 1966), 6.

Philosophy, which had begun in 1797.[1] Both these publications included original articles and letters, papers reprinted from other journals, translations of papers from foreign languages, book reviews, and reports of the meetings of scientific societies at home and abroad. Benjamin Silliman's *American Journal of Science*, which appeared from 1818, followed the same general plan. These journals were valuable to amateurs, who in their pages appeared on a level with more expert authors with whom they could therefore engage in occasionally fruitful dialogue. To established men of science—to talk of professionals would still be an anachronism—such journals were useful because they gave rapid publication to informal communications. *Nicholson's Journal* and the *Philosophical Magazine* were joined in 1813 by Thomas Thomson's *Annals of Philosophy*; unlike the other two editors, Thomson was an academic scientist who had written an excellent textbook of chemistry, was one of the major contributors to the early editions of the *Encyclopedia Britannica*, and went on to become professor of chemistry at Glasgow. Nicholson had translated chemical works from the French—the language of Lavoisier—had produced a *Dictionary of Chemistry*, and had performed some experiments in electrochemistry; while Tilloch was a journalist. The length of run of their journals was in inverse proportion to the scientific expertise of the editors: the *Philosophical Magazine* is still with us; having absorbed *Annals of Philosophy* after a run of fourteen years, and Nicholson's *Journal* after it had been going for nearly eighteen years. What is surprising is that between 1798 and 1826 there were at any time in Britain two general scientific journals, which included papers of a high order of importance; claims that this was a period when science was in decline should clearly be treated with caution.

It is to these journals that the student looking for the norm in

[1] A. Ferguson (ed.), *Natural Philosophy through the 18th century* (London 1972) —reprints papers which first appeared in 1948; M. P. Crosland, 'Humphry Davy—an alleged case of suppressed publication', *British Journal for the History of Science*, 6 (1973), 304–10.

early nineteenth-century science in Britain or America should turn first. He will find that chemistry and natural history, including geology, predominate in their pages; and that there is an interesting mixture of contributors, some names being familiar to every student of science, while others are those of country squires, travellers or cranks. It is clearly impossible to apply the categories we use to classify scientists to the scientific community of the early nineteenth century. During the nineteenth century, and particularly after 1826 when it had absorbed *Annals of Philosophy*, the *Philosophical Magazine* followed in the path of the Royal Society's *Philosophical Transactions* and became increasingly technical and prestigious, concentrating upon physics; so that by the end of our period, when it was publishing the papers of J. J. Thomson and Ernest Rutherford, it had long ceased to be a general scientific journal. *The Edinburgh Philosophical Journal*, which ran in various series from 1819 to 1864 under the aegis of the experimentalist and devoted Newtonian Sir David Brewster, can perhaps be described as a general journal; but it seems that in the middle of the century there was a gap which was not filled until 1869 when Macmillan's launched *Nature* under the editorship of Norman Lockyer,[1] a civil servant with a passionate interest in astrophysics and a determination to get on in the world. By 1869 there was less scope for the dilettante in science, and the refereeing for *Nature* was a good deal stricter than it seems to have been for the early *Philosophical Transactions* or *Philosophical Magazine*; but the mixture of articles, (some reprinted, some translated, and some original), reviews and reports was not very different, and the correspondence section provided an outlet for amateurs or for argumentative professionals.

Nature printed presidential addresses from the British Association meetings in the week in which they were delivered, thus anticipating by months their official publication in the *Report* of the British Association for the Advancement of Science. Other societies were not so generous; the Royal Society, for instance, would not publish any paper in their *Philosophical Transactions*

[1] On Lockyer, see A. J. Meadows, *Science and Controversy* (London 1972).

which had already appeared elsewhere.[1] Even an abstract in a foreign language of a paper delivered before them could cause difficulties. Authors had to choose whether to get their results known quickly in a less august publication, or to wait for the *Philosophical Transactions*; and sometimes they had to choose whether to append scientific observations relating to a voyage to a published narrative of the voyage, or to wait until the Royal Society would publish it later. During the nineteenth century, the Society's own *Proceedings* developed from a mere abstract of the *Philosophical Transactions* into a less-formal journal in its own right; but its transition into a specialised publication, for the most part unintelligible to the layman, was rapid. The *Report* of the British Association has already been mentioned; the Association was founded at York in 1831 and has met each year since in various provincial cities, and occasionally overseas. Its foundation can be attributed in part to emulation of the Germans, who had had a similar body for some years and whose science had been invigorated thereby; partly to dissatisfaction with the Royal Society which was felt to be too much connected with what we would now call the Establishment; and partly to an assertion of the provincial spirit against the assumption of Londoners that everything of importance happened in the metropolis. The meetings were open to the public; there were various sections devoted to different sciences, of which in the early days geology was the most popular.

The meetings of the British Association[2] became one of the great events in the diaries of scientists, and distinguished figures

[1] See L. P. Williams, *Selected Correspondence of Michael Faraday*, 2 vols. (Cambridge 1971), I, 480–1, 487; J. Ross, *Voyage of Discovery* (London 1819), appendix I (on magnetic deviation). There was no objection to publishing elsewhere afterwards; see indeed C. Babbage, *The Decline of Science in England* (London 1830), 105.

[2] G. Basalla, W. Coleman, R. H. Kargon (ed.), *Victorian Science* (New York 1970); J. B. Morrell, 'Individualism and the Structure of British Science in 1830', *Historical Studies in the Physical Sciences, 3* (1971), 183–204; A. Thackray, *John Dalton* (Cambridge, Mass. 1972); A. D. Orange, 'Origins of the British Association', *British Journal for the History of Science, 6* (1972), 152–76.

such as Liebig and Helmholtz were invited to the meetings as well as scientists from all over Britain. Large numbers of papers were read; the bulk of these were never printed, for only abstracts appeared in the published *Report* of the meetings. Bibliographically the *Reports* are tiresome because the first two meetings were reported in one volume, and subsequent meetings in one volume each; so that meeting numbers and volume numbers do not agree. Victorians anyway usually referred to the meetings by their place—York, Oxford or Aberdeen for example—rather than by year or number; but each volume gives the places of all former meetings, with names of the principal officers on each occasion. More tiresomely, in the earlier volumes the abstracts of papers were separately paginated from the addresses, reports of committees, and so on; so that in a given volume there would be for instance two page 89s.

Despite this rather chaotic arrangement, which might daunt the historian, there is no richer mine of information about nineteenth-century science than these *Reports*. The Association set up committees to discuss and report on questions felt to be important; it numbered among its officers, who held office only for a year, men whose names have come down as familiar to posterity, and others who are forgotten but must not be neglected by the historian—because to hold office in the Association was an indication of importance in the scientific community. Efforts were usually made to get a local man, distinguished as a scientist or noted for his interest in science, as President. He always gave a long address to the meeting; these were usually carefully prepared over a long period, as one can see from the correspondence of various presidents. Usually they reviewed the progress of science generally over the previous year, and probably that of their own particular science over a longer period; for no branch of science monopolised the presidency. If they were working scientists, they might give an account of the problems upon which they were engaged; otherwise they might give warnings to government of the need to spend more money on scientific education or research, and to industry of the threat from Germany or

America if scientific methods of research and quality-control were neglected.

Each year, then, the President was the spokesman for the scientific community. Sometimes he chose to disseminate views which were simply his own, or those of a small group; as when in 1874 in Belfast John Tyndall urged his audience to embrace an agnostic materialism. This caused a great scandal; and in general the presidential addresses are important for their representative character rather than for any particular originality, and were intended to be so. In the 1850s the Presidents of the various sections also began to give addresses which were printed in the *Report*, and also often—like the President's—elsewhere. These addresses were usually, but not always, shorter and more concerned with internal problems in the various sciences; a review of the work done in the previous year is a general feature of them. These help us to see where the frontiers between the various sciences were drawn at a given period, and to trace the dissemination of ideas across frontiers; thus it was in his presidential address to the Mathematics and Physics Section in 1888 that George Fitzgerald drew the attention of British scientists to the experiments of Heinrich Hertz on electromagnetic waves which paved the way for radio. Thus while the British Association's *Report* is not a general journal in the usual sense, we can from the presidential addresses, the reports of its committees, the lists of officers and members, the abstracts of papers read, and even the accounts, learn a great deal about the way science was practised in the Victorian period. We see various pressure-groups at work, and we see scientists becoming an increasingly self-conscious and professional group; and we also follow the development of the various sciences, and the spread of new ideas within them. For this period the British Association can help us to recognise a norm, as the Royal Society with its closed membership of increasingly eminent scientists no longer does after the 1820s.

The British Association drew upon provincial pride; but there were in provincial cities scientific societies whose published transactions sometimes take the form of a general scientific journal; and whatever form they may take, they enable us to assess the state of science in the city. Most of these societies were concerned chiefly with antiquities or with natural history; the most famous, and one which covered the whole range of science, was the Manchester Literary and Philosophical Society.[1] The eminence of John Dalton, the founder of chemical atomic theory, and of James Prescott Joule, a great name in the early history of thermodynamics, has perhaps diverted attention from the representative and therefore amateurish nature of the society. Its journal does not contain an endless series of difficult papers pushing back the frontiers of knowledge; such papers are there occasionally, but the student searching should expect to find not these, but papers giving a good idea of what counted as science in Manchester in the nineteenth century.

Among general journals there were not only regional differences, but also differences in the social class of the readership. We are often told, and it must be true, that Victorian Britain lagged behind Germany and America in the training of technicians; but there were enough of them to support scientific journals explicitly directed at 'mechanics'. Benjamin Martin, an instrument-maker and a tireless Newtonian populariser, produced his *General Magazine of Arts and Sciences* in fourteen volumes between 1755 and 1765; but this was really a kind of encyclopedia published in parts rather than a journal. The *Mechanics Magazine* began in 1823, at the very outset of that 'March of Mind' which delighted Jeremy Bentham's disciples; it was hardly an inflammatory publication. It ran for fifty years; but for the latter part of the century (from 1865) there was the *English Mechanic* which in columns of small type included an amazing amount of information, both about developments in science and about technical progress. One can find in its pages, for example, an amateur astronomer reporting on his discovery of some faint and distant

[1] D. S. L. Cardwell (ed.), *John Dalton and the Progress of Science* (Manchester 1968), 1–10; on Dalton and Joule, see W. L. Scott, *The Conflict between Atomism and Conservation Theory* (London 1970). Arnold Thackray is studying the Manchester Literary and Philosophical Society.

star; and instructions, spread over several issues, for making a motor-bicycle or a phonograph. The sublime science of astronomy seems indeed to have retained its prestige ever since the days of Copernicus; perhaps particularly in the English-speaking world, where it was associated with the great name of Newton.[1] In the nineteenth century the patient and industrious amateur equipped with a telescope could hope to make genuine discoveries, though probably not ones of great general interest, in astronomy. He might well report them in a mechanics' journal;[2] and it is a pity that the *Royal Society Catalogue* did not descend to this scientific literature. But of course it is to the student of the dissemination of science and the social history of science that these journals will be of most value; he will be sorry that many of the contributions took the form of letters which were not indexed, and that in general the indexing leaves much to be desired. There is no remedy except to go through patiently.

The Royal Institution in London seems to have begun with a dual emphasis, some founders like Count Rumford and Sir Thomas Bernard (of the Society for Bettering the Condition of the Poor) having in mind a technical college, while the majority were 'improving landlords' hoping for the benefits of applied science. These they duly obtained in the researches of Humphry Davy on tanning, agricultural chemistry and the safety-lamp for coal mines; scientific lectures at a high level were also delivered there, to a fashionable audience—few artisans succeeded in getting in. In 1802–03 a *Journal* was published at the Royal Institution, consisting mostly of reviews and abstracts of papers from other publications. From 1816 to 1836, a general quarterly *Journal of Science and Arts* was published under the editorship of

[1] P. Miller, *The Life of the Mind in America* (London 1966), 278; H. Davy, *Collected Works*, ed. J. Davy, 9 vols. (London 1839–40), viii, 326.

[2] For example the Rev. T. H. E. C. Espin, of Wolsingham Observatory, Co. Durham, did this; he was eminent enough to have a crater on the back of the Moon named after him recently, and to have received a medal from the Royal Astronomical Society in 1913.

W. T. Brande, who had succeeded Davy as Professor of Chemistry; this contained reviews and original articles and is a useful source in the same way as are the *Philosophical Magazine* and the *Annals of Philosophy* for the same period. After a lapse of twenty years, the Royal Institution then in 1851 began to publish *Proceedings*; these contained at first abstracts, but then full texts, of the discourses delivered each Friday evening during the season. For the end part of our period these are very valuable material.[1] They were delivered by eminent scientists, describing recent work in a field for a lay audience; often therefore one had a discoverer being his own populariser, though sometimes other men's work was the main subject. We thus learn both about science and about its dissemination; and we meet the giants of Victorian science describing their researches and reviewing those of others, in a less formal way than in an ordinary scientific paper. The chief work of the Institution was in physics and chemistry; and the majority of the discourses can be broadly classified as belonging to these disciplines. But there were also many on astronomy, geological sciences and other topics. From these various publications, we can learn much about what was one of the most distinguished centres of advanced scientific research and of scientific lecturing in nineteenth-century England, and indeed in the world.

There was another *Quarterly Journal of Science* which is sometimes confused with that published at the Royal Institution; but this was a private journal, of a general kind but with an emphasis on physical science, which appeared between 1864 and 1885. William Crookes the chemist was its editor; and it is worth looking at although it seems never to have been very influential, lacking the wide readership of *Nature* and the prestige of the *Philosophical Transactions.* The only English scientific journal which set out to pay all its contributors was *The Laboratory*, which was chiefly devoted to chemistry. It began in 1871, and

[1] They are being republished as the *Royal Institution Library of Science*, classified under *Physical Sciences* (11 vols., London 1970), *Earth Sciences* (3 vols., 1971), and *Astronomy* (2 vols., 1970).

died in the same year. The British Association *Reports*, the journals intended for mechanics, and the Royal Institution's publications all dealt in various degrees with technology—then called 'arts'—as well as pure science; indeed the two categories then as now are often impossible to separate. But the Royal Society of Arts[1] published *Transactions* from 1783 to 1848 which are entirely concerned with applied science. The society awarded premiums and medals for new devices or improvements; it indicates the determination of Englishmen of the late eighteenth century to turn science to account in agriculture or in manufactures. Its heyday coincided with the reign of Sir Joseph Banks, the landowner, explorer and natural historian, as President of the Royal Society from 1778 to 1820; a period in which science came to be taken seriously, rather than as something a polished person ought to know a little about, by those in powerful positions in Britain. The practical emphasis of the Royal Society of Arts and its *Transactions* enables us to see why this happened and to identify some of the patrons of innovation. Covering both arts and sciences, there was an abstracting journal, *Retrospect of Discoveries* which came out between 1805 and 1815; it had not much authority, but it is upon occasion useful.

This discussion of general journals has been almost confined to those published in Britain; but the pattern of publications in other countries seems to have been not dissimiliar. In France the Academy of Sciences published a journal of even greater dignity and slowness than the Royal Society's;[2] in the eighteenth century there too private journals began offering more rapid publication and more news from abroad, the best known being François Rozier's *Observations sur la physique*. Sometimes these journals

[1] D. C. G. Allan, *William Shipley: founder of the Royal Society of Arts* (London 1968); D. Hudson and K. W. Luckhurst, *The Royal Society of Arts, 1754–1954* (London 1954).

[2] R. Hahn, *The Anatomy of a Scientific Institution: the Paris Academy of Sciences, 1666–1803* (Berkeley 1971); C. Babbage, *Decline of Science in England* (London 1830), 190–2; H. Guerlac, *Lavoisier—the crucial year* (Ithaca, New York 1961); and Crosland, see note 1, p. 107.

stood on opposite sides in some scientific schism; thus in the early nineteenth century the *Annales de Chimie* welcomed papers written in terms of Lavoisier's chemical theories, while the *Annales de Physique* clung to the older views. In Germany, the *Miscellanea Curiosa* began to appear in Leipzig in 1670; and the *Acta Eruditorum* in which Leibniz was the greatest luminary, in 1682. The academies of Berlin and of St Petersburg followed the example of that of Paris in publishing their transactions as in other things.

All over Europe provincial societies published papers which achieved a more or less wide circulation. Thus one of the seminal papers of the nineteenth century, that by Gregor Mendel on inheritance,[1] appeared in the *Transactions* of the Natural History Society of Brünn (now Brno in Czechoslovakia) in 1866. It remained unnoticed for more than thirty years; but this was not simply because few of those who belonged to the Society would have been able to see its implications or follow them, for the journal went to the major academies and learned societies of Europe, and Mendel was in correspondence with some prominent biologists who also failed to see the fundamental importance of his researches. By way of contrast, Willard Gibbs' papers on thermodynamics appeared equally obscurely in the *Transactions* of the Connecticut Academy of Sciences;[2] but he sent offprints to Clerk Maxwell in Britain and to Wilhelm Ostwald in Germany, and they did see how important his work was. The situation in America down to the end of our period seems rather different from that in Europe; American science was in general

[1] R. Olby, *The Origins of Mendelism* (London 1966); G. Mendel, *Experiments on Plant Hybridisation*, ed. J. H. Bennett (Edinburgh 1965); C. Stern and E. R. Sherwood, *The Origins of Genetics: a Mendel Sourcebook* (San Francisco 1966).

[2] N. Reingold, *Science in 19th-century America* (London 1966), 315–22; see also Reingold's edition of *The Papers of Joseph Henry* (Washington, D. C. 1972–); B. Z. Jones (ed.), *The Golden Age of Science* (New York 1966); R. P. Stearns, *Science in the British Colonies of America* (Urbana, Ill. 1970). Science and technology also appear in H. M. Jones, *O Strange New World. American Culture: the formative years* (London 1965).

provincial compared to that of Europe, but it was further provincial in that there was no great metropolis like London, Paris or Berlin to attract or repel the most eminent of scientists. In Boston, New York and Philadelphia there were scientific societies of some note whose journals are interesting though they naturally contain few papers of the first rank. The Smithsonian Institution, in Washington D.C., was set up in 1846, with Joseph Henry, whose work on electromagnetism had given him an international reputation, as its secretary and first director. From 1847, the Smithsonian began publishing *Annual Reports*; and from the following year *Contributions to Knowledge*. The *Annual Reports* contain obituaries of scientists, both American and foreign—sometimes reprinted or translated from other journals—which are useful; and both journals are valuable, like Silliman's *American Journal of Science* which has already been mentioned, for illuminating the development of science in nineteenth-century America and the role which it played in the life of the nation.

These general journals give one a good picture of the context of science; but in the nineteenth century there began an epoch of specialisation. Increasingly as the century went on, chemists, biologists and astronomers came to publish in different journals. By the end of the century, botanists and zoologists were usually publishing in their own journals; and chemists were beginning to divide themselves into 'organic', 'physical' or 'inorganic', publishing in different journals or different sections of the same journal. Such great generalisations as the Principle of Conservation of Energy and the Theory of Evolution linked together sciences which had previously seemed distinct; but at the more mundane level the sciences became more specialised in vocabulary and techniques, and grew apart. This process happened at different rates in different countries; it depended upon the professionalisation of scientists, and therefore upon the state of scientific education and the demand for specific kinds of scientists in government and in industry. The working of legislation on pure food and drugs, for example, required chemical analysts, and the dye industry organic chemists: by the 1870s in England the tensions between these experts in applied science and the academic chemists were such that they formed different societies which have only in recent years united themselves.[1]

But long before this, devotees of the most popular sciences had formed themselves into societies which usually published their own journals. Thus in England, as one might expect, it was the natural historians who first formed their own society, the Linnean Society, which conserved the collections made by the great classifier Linnaeus, bought after his death by Sir James Smith.[2] The Linnaean Society's *Transactions* appeared from 1791; it was there in 1858 that the theory of evolution by natural selection was first proposed, by Darwin and A. R. Wallace. Nobody took any notice of so original and theoretical a contribution; in general the pages of this journal were filled with sound papers essential to anybody working on the internal history of biology in the nineteenth century.

The geologists were the next to organise themselves; and the Geological Society of London began bringing out its *Transactions* in 1811.[3] Right through the nineteenth century, geology was a science which raised general questions of importance to miners, theologians, philosophers, physicists and chemists; to write its internal history would be to tell only half the truth. But the *Transactions* were intended to be a record of facts rather than a

[1] The two institutions were the Chemical Society of London, and the Royal Institute of Chemistry; the tensions can be seen in *Chemical News* in the 1860s and 1870s; see, on professionalisation in chemistry, B. W. G. Holt, in the *British Journal of Sociology*, *31* (1970), 181ff. On the unification of science, see H. Sharlin, *The Convergent Century* (New York 1966); Y. Elkana, *Discovery of the Conservation of Energy* (London 1973). On specialised journals of the twentieth century, see P. Brown and G. B. Stratton, *World List of Scientific Periodicals, 1900–1960*, 3 vols., 4th ed. (London 1963–65); D. J. Grogan, *Science and Technology: an introduction to the Literature* (London 1970).

[2] F. A. Stafleu, *Linnaeus and the Linnaeans* (Utrecht 1971); W. Blunt, *The Compleat Naturalist* (London 1971), 236–8; C. Linnaeus, *A Tour in Lapland*, ed. J. E. Smith, 2 vols. (London 1811), I, vii–x.

[3] H. B. Woodward, *History of the Geological Society of London* (London 1907); M. Rudwick, *The Meaning of Fossils* (Cambridge 1973); and C. C. Gillispie, *Genesis and Geology* (Cambridge, Mass. 1951).

collection of speculative papers; and this intention was almost too fully realised. Through the *Transactions* one can trace the development of the science from the early years, when the descriptions were mostly mineralogical and the tone that of the County natural histories in the tradition of Robert Plot's *Oxfordshire*, down to the period when the rise of palaeontology made the dating of rocks a possibility, and the coming of evolutionary theory integrated biology with geology.

Sir Joseph Banks, as President of the Royal Society, had welcomed the setting up of the Linnaean Society; but he became increasingly worried that the development of specialised societies would lead to the weakening of the Royal Society. Nevertheless, the Geological Society was followed soon after Banks' death in 1820 by the Royal Astronomical Society[1] whose *Memoirs* began to appear in 1825, and *Monthly Notices* in 1827. Soon afterwards came the founding of the Royal Geographical Society. Chemistry had been in an exciting transitional stage in the first decade of the nineteenth century and was correspondingly popular; but it then entered upon a phase of consolidation, and the Chemical Society of London was not founded until 1841, its *Journal* beginning publication in 1848. Like the other specialised societies, its publications at first concentrated heavily upon the facts rather than on their interpretation; indeed it was the policy of the *Journal*, through most of the nineteenth century, not to accept papers of a purely theoretical nature.

As the general journals published by societies had been complemented by private journals, so the specialised societies' publications were soon joined by private journals covering the same field. The *Botanical Magazine*, with superb plates which made it and still make it a collectors' piece,[2] began in 1787; it was a

[1] J. L. E. Dreyer and H. H. Turner, *History of the Royal Astronomical Society, 1820–1920* (London 1923); C. R. Markham, *The Fifty Years Work of the Royal Geographical Society* (London 1881); Chemical Society, *The Jubilee . . . 1841–1891* (London 1896); T. S. Moore and J. C. Philip, *The Chemical Society, 1841–1941* (London 1947).

[2] See W. Blunt, *The Art of Botanical Illustration* (London 1950), chap. XV;

serious contribution to botany as well as a source of beautiful plates, particularly of exotic plants which it was the chief intention of the journal to describe. In 1838 there began to appear the *Annals of Natural History*; and the *Gardener's Chronicle* was a valuable source even for Darwin, and shows the lack of any gulf during much of the nineteenth century between the amateur pottering in his garden and the botanist. In chemistry there were various private journals, most of which ran for only a brief period until William Crookes began *Chemical News* in 1860. This had a format very like that which Lockyer was to use for *Nature*, and it appeared weekly as *Nature* was to do. Crookes was a speculative man, and he encouraged theoretical and hypothetical papers in his journal, which also carried correspondence and is very illuminating on both the internal and external history of chemistry in the latter part of the nineteenth century.

It is in these more specialised journals, whether private or run by a society, that it is perhaps easiest to follow the progress of science; usually a dialectical progress, or at least a zig-zag rather than a steady advance of knowledge. A difficulty is, that often opponents in some controversy did not send their papers to the same journals, so that one gets a curious effect of shadow-boxing as one goes through one journal. Surprisingly frequently, they were not even aware of one another; thus in nineteenth-century chemistry,[1] scientists in different countries, or even different parts of one country, seem to have worked in more or less complete ignorance of what was going on elsewhere. Work that seems to us clearly relevant to their concerns they were either ignorant of, or failed to remark. The most important innovators were often those who had spent some time abroad and had therefore gained access to a foreign tradition; the synthesis of this with

the various series of the various journals are listed in my *Natural Science Books in English, 1600–1900* (London 1972), 227ff., except for the *Gardener's Chronicle*, *1* (1841–); and see H. R. Fletcher, *The Royal Horticultural Society, 1804–1968* (London 1969), 158.

---

[1] C. A. Russell, *The History of Valency* (Leicester 1971); and Scott, see note 1, p. 112, discusses this question.

their native tradition led to rapid advance and the clearing up of difficulties which had halted real advance in both the countries concerned.

Indeed, ease of scientific communication across national and linguistic frontiers is something that we tend to take too much for granted; it presents a rich field for investigation by the historian, and the study of specialised journals can help him along.[1] By the nineteenth century it had become accepted that one should in a learned article refer to other writings which had proved valuable. That a man does not cite a paper is no proof that he had not read it and perhaps even copied from it; nor is a citation proof that he had read it, for the copying of somebody else's footnotes is an old malpractice: but citations are evidence, even if they are not cast-iron evidence. That somebody translated a paper is of course evidence that he read it, and that he saw it as important. Nineteenth-century British scientists did not on the whole find it easy to read or speak modern foreign languages; German especially was a stumbling-block until the latter part of the century when it became essential for any aspiring chemist to do his Ph.D. at a German university. Before that time, Humphry Davy, Michael Faraday, J. D. Hooker and Charles Darwin were among those who had tried and failed to master the language. French was easier; and indeed during the Revolutionary and Napoleonic wars, scientific communication between Britain and France seems to have suffered very little although relations with Germany and Sweden became much more difficult.

Various journals did publish translations of foreign papers, and curiously enough the *Philosophical Transactions* included in 1800 Volta's fundamental paper on electricity in French; it was translated in the *Philosophical Magazine*. In the middle years of the century there was one English journal which devoted itself entirely to translations of papers which had appeared in foreign publications; Richard Taylor's *Scientific Memoirs*, which was published between 1837 and 1853. These were full translations,

[1] See J. Z. Fullmer, *Sir Humphry Davy's Published Works* (Cambridge, Mass. 1969), 2–5.

unlike some of those which appeared in other journals; they were also made directly from the original language, in contrast to some 'translations' from the German in the early nineteenth century which were in fact prepared from a French version, itself often an abstract or paraphrase. In early translations one sometimes finds curious effects; the work on chemical affinity which C. L. Berthollet read to the short-lived academy set up in Egypt on Napoleon's expedition there was translated in various numbers of the *Philosophical Magazine*.[1] The first parts of it were heavily cut, while the later parts are essentially a complete translation. Long papers like this, one might notice, appeared in parts over several numbers of the journal, and it might be about a year between the beginning and end of a paper; right through the nineteenth century our ancestors read their science, like their novels, in parts.

Translations, and reprinting of articles from other journals, provide evidence of the reputation of an author; particularly when taken with election to important societies or academies at home and abroad. They are not necessarily a guide to a man's reputation outside the immediate community of specialists in his science; for this we have another indicator in the *Reviews* which were such a feature of nineteenth-century intellectual life.[2] The formidable *Edinburgh Review* discussed works of science, including even lectures read before the Royal Society, bringing out the points involved for a lay audience. Usually these were not merely scientific points; thus reviewing the work of Laplace provided an opportunity for an assault upon the English universities for their backwardness in mathematics, while contrariwise a review of Thomas Young's theory of light became an occasion

[1] C. L. Berthollet, *Philosophical Magazine*, *9* (1801), 146–53, 342–52; *10* (1801), 69–74, 129–42; another version was *Researches into the Laws of Chemical Affinity*, tr. M. Farrell (London 1804). On papers translated from German to English, see B. S. Gower, 'Speculation in Physics . . . *Naturphilosophie*,' *Studies in the History and Philosophy of Science*, *3* (1973), 301–56.

[2] W. E. Houghton (ed.), *The Wellesley Index to Victorian Periodicals, 1824–1900*, 2 vols. (Toronto 1966–72).

for denouncing a heretic from Newtonian orthodoxy. The other *Reviews*, the *Quarterly*, the *Westminster* and the *North British*, for example, followed the *Edinburgh* in devoting a considerable amount of space to the discussion of works of science through most of the century; and the *Nineteenth Century* carried articles by eminent scientific men drawing attention to current science. What they reviewed or emphasised gives us a valuable clue to what seemed important or interesting at the time; and when the review was written, as it often was, by an eminent scientist rather than by a literary hack it was often used as an opportunity for taking stock, for looking at the wood rather than the trees. The authors of most of the articles in the major reviews have been identified; the historian of nineteenth-century Britain owes a debt of gratitude to the compilers of the *Wellesley Index*.

The *Edinburgh Review* had some predecessors, such as the *Monthly Review*, which had also looked at scientific publications; but had lacked the authority, real or assumed, of its nineteenth-century successors. Such other fascinating publications as the *Gentleman's Magazine* and the *Ladies' Diary* carried scientific material; the mathematical problems set in the latter publication provided John Dalton with his first scientific exercise, and the pages of the *Gentleman's Magazine* contain not only obituaries of scientists who were not elsewhere written up, but also contributions—usually amateurish—to scientific debates; as for example to electrochemistry in the opening years of the nineteenth century. Newspapers, and weekly publications such as *The Athenaeum*, sometimes give us reports of meetings which are not elsewhere described in print, or of events at meetings which were not mentioned in the official reports. These may not be in any way scandalous; thus at the British Association in 1851 the celebrated French chemist J. B. Dumas gave an important but informal address on the arrangement of the chemical elements, of which the account in *The Athenaeum* (p. 750) is the fullest we have because the *British Association Report* confined itself to official matter and to abstracts of formal papers.

In Britain, therefore, and *mutatis mutandis* in other countries also, we find that journals more or less devoted to science cover a wide spectrum: from the austere papers of Newton or Faraday through to propaganda and gossip. The careful study of the more specialised journals will show us minds gaining ground upon the dark, and illuminate for us science as an intellectual activity in which firmly established facts are brought into a coherent framework by theory, which in turn generally leads to the discovery of more facts. It will indicate to us who read, or claimed to have read, which authors and publications; and by following citations we can work our own way back into the science of the period, although such working backwards has its problems for the historian.

Even the most austere publications have much to tell us also about the workings of the scientific community; we can find out about editorial policy and refereeing, and see whose papers got in and whose did not. Thus in the mid-nineteenth century it seems that, as one would expect, only really established scientists could get speculative papers on atomic theory published in the most famous journals;[1] and as we suggested above, it was no doubt easier for one who had been Secretary and acting President of the Royal Society to publish in its *Philosophical Transactions* a paper on the unserious-seeming topic of why the eyes in portraits seem to follow one around, as W. H. Wollaston did in 1824. One may also find, as it has been alleged that one can in the case of Jean d'Alembert, evidence of 'premature' publication of half-baked material by an ambitious scientist anxious to forestall competitors. To establish a claim of this kind one would want to know that a man could do better; and preferably that there was some occasion—election to an academy for

[1] On publication of papers on atomic theory, see S. G. Brush, 'The Kinetic Theory of Gases', *Annals of Science*, *13* (1957), 188–98, 273–82; *14* (1958), 185–196, 243–55; G. R. Talbot and A. J. Pacey, 'Some early kinetic theories', *British Journal for the History of Science*, *3* (1966), 133–49; and the introduction to J. B. S. Haldane (ed.), *Collected Scientific Papers of J. J. Waterston* (London 1928); T. L. Hankins, *Jean d'Alembert* (Oxford 1970), 31, 42ff., 50–1. See note 1, p. 106.

example—for which he felt that he must publish or perish.

From specialised journals we can also learn something about editorial policy which might be collectively enforced—at least in principle—as in the publications of scientific societies, or idiosyncratic as it could be in private journals; though the editor of these always had to keep an eye on his circulation figures.[1] To know where a paper appeared is, in assessing its importance, as vital as to know when it appeared. Editors of journals clearly played a very important part in the history of science, and yet they are often much more shadowy figures than their contributors, in the present state of knowledge. One can try to quantify some studies of journals; thus the number of journals devoted to some specific science may be an indicator of its importance at the time, and so may the proportion of space in general journals allotted to the various sciences. This seems a straightforward matter to determine; but given the fluid frontiers of the sciences it is by no means easy to determine in all cases in which category a paper should be put. We may, for example, find in a journal ostensibly —and genuinely—devoted to chemistry great numbers of papers which we would classify as mineralogy, physiology, or electricity. A journal may retain its title even when in the course of time its contents have changed; in any quantitative study, whether of members of a society or of journals, we must beware of bogus precision, while accepting that, like statistical studies in medicine, quantitative investigation in this field may define problems to be solved by other means. There is, no doubt, some correlation between the rate of progress in a science and the energy put into

[1] On the problem for contemporaries in classifying original scientists and their papers, see T. S. Kuhn 'Energy Conservation', in M. Clagett (ed.), *Critical Problems in the History of Science* (Madison, Wis. 1959), 321–56; D. S. L. Cardwell, *From Watt to Clausius* (London 1971), 213ff. to esp. 231; F. Szabadváry, *History of Analytical Chemistry*, tr. G. Svehla (Oxford 1966), 241; F. Greenaway, 'A pattern of chemistry', *Chemistry in Britain*, 5 (1969), 97–9; J. W. van Spronsen, *The Periodic System of Chemical Elements* (Amsterdam 1969). These two last refer to the Periodic Table classification in chemistry; many of the original papers relating to this are reprinted in my *Classical Scientific Papers, Chemistry: 2nd series* (London 1970).

it, which is perhaps correlated again with the number of men and journals devoted to it.

The more general journals can tell us even more about these points, and about the organisation of science and the order of prestige among the sciences at various times. In journals as in manuscript materials, we see scientists advancing knowledge, often disputatiously and self-assertively; and engaged in struggles for more recognition and support for science generally, and for their own science in particular. That is, we see them as people in particular situations, and not merely as names to which particular chemical reactions or physical constants are attached. In the next chapter we must look at books which throughout our period were the chief vehicle both for the radically new ways of looking at phenomena that we associate with scientific revolutions, and, as textbooks, for passing on the received view.

CHAPTER 5

Scientific Books

To most historians of science, trained initially in philosophy or in modern science, books of a formal or mathematical kind usually seem less recondite than works of descriptive science. Voluminous journals of travels, and long descriptions of species and minerals, or of miscellaneous chemical reactions, do not appear to raise general questions or to be related to important aspects of modern science; they therefore tend to be relegated to the antiquarian. In the twentieth century we have come to expect that science will be formal and quantitative, and that 'the natural history stage' is an awkward period, a kind of puberty, in the history of each science. Descriptive science thus seems unworthy of general or prolonged attention; and its chief aim, classification, seems a dull and mechanical business. This is curious for several reasons; and one may hope that, as historians of science trained as historians increase in number, this condescending approach to the dominant science of most of our period will be given up. Descriptive science received government support; and it formed the programme of such men as Linnaeus, Buffon, Goethe, and Sir Joseph Banks.[1] Applied mathematics also had its devotees and enjoyed great prestige, particularly in eighteenth-century France; but to most contemporaries it seemed abstruse and inaccessible, and certainly not the paradigm for the whole of science.

The problem of classification, of family resemblances and natural kinds, had faced philosophers and scientists at least from Aristotle to Darwin, and in our century has been discussed by

[1] H. B. Nisbet, *Goethe and the Scientific Tradition* (London 1972); F. A. Stafleu, *Linnaeus and the Linnaeans* (Utrecht 1971), discusses taxonomy; as does P. C. Ritterbush, *Overtures to Biology* (New Haven 1964), esp. chap. IV. See T. L. Hankins, *Jean d'Alembert* (Oxford 1970), 92ff. for the mathematics/natural history division.

Wittgenstein. Those who feel that modern nominalism has disposed of the problems of taxonomy, and that a goose might as reasonably be put among the *edibilia* as among the *anatidae*, thereby disqualify themselves from the serious study of much past science. In recovering the picture of science as contemporaries saw and used it, we must allow to natural history and geography a prominent place in the foreground; and the historian of science must not, particularly in discussing books as sources, underrate natural history, of which the literature is much more voluminous than that of the physical sciences. In agricultural science, natural history passes into technology; and the literature of technology, too, is something which the historian of science must not neglect.

In devoting two chapters chiefly to books we are indicating their enormous importance as sources. The scientific journal has indeed become increasingly important through our period, being the vehicle for 'normal science' and an invaluable means of communication within the scientific community. But most people have always learned their science not from journals but from books, and this has become more necessary as journals have become increasingly specialised. Textbooks have sometimes been written to impress upon the rising generation a different view of a whole science, as for example was Lavoisier's *Elements of Chemistry*;[1] but this happens only when an innovator decides as it were to popularise himself, and most textbooks provide a received view rather than a thoroughgoing reinterpretation of a science. But scientific revolutions have usually been brought about by books rather than by papers, because in a book the author has had room to marshall his arguments; and similarly systematic descriptions of floras and faunas, investigations of geological phenomena or of the distribution of species, and accounts of scientific expeditions,

[1] A. L. Lavoisier, *Elements of Chemistry*, tr. R. Kerr, 1790—reprint, ed. D. McKie (New York 1965); M. P. Crosland, *Historical Studies in the Language of Chemistry* (London 1962), discusses many standard works; and on the author of a very successful textbook, see W. A. Smeaton, *Fourcroy, 1755–1809* (Cambridge 1962).

have needed more space than a journal can provide. The temptation to concentrate upon great books must be resisted; the historian's primary task is to recover the norm, and it is only against this norm that books can be assessed. It is therefore sensible to read a wide range of books from a relatively narrow span of time. Any distinction between sources for the internal history of science and those valuable for background is hard to make; but for convenience we shall in this chapter look at what are clearly scientific books, and in the next at works which are related to the history of science but were not themselves regarded as contributions to science.

The place to begin is with bibliographies and library catalogues; booksellers' catalogues can also be very useful. The bibliographer's interests are not the same as those of the historian, but if the two proceed in ignorance of each other they will both be losers. Bibliographers can often find problems fascinating to them in works of science, particularly in the field of natural history where books often appeared in parts over a number of years, and where the plates were reworked and may be found in a number of states.[1] The practice of publication in parts means that one must beware of taking too seriously the date on the title-page of a book; most of it may have appeared before or after this date, and the wrappers in which the parts originally came out will probably have been destroyed when it was bound up. This is something the historian must beware of before he concludes that 'X could not have read Y's work because it did not appear until the following year'. There is a Society for the Bibliography of Natural History, and the historian is well advised to browse in its *Journal*.

[1] See G. Grigson and H. Buchanan (ed.), *Thornton's Temple of Flora* (London 1951); S. Sitwell, H. Buchanan and J. Fisher, *Fine Bird Books* (London 1953); S. Sitwell, W. Blunt and P. M. Synge, *Great Flower Books* (London 1956); A. M. Coats, *The Book of Flowers* (London 1973). On the use of different plates for different editions, see E. G. Turbott, *Buller's Birds of New Zealand* (London 1967); the reprint of J. V. Thompson, *Zoological Researches, 1824–1834* (London 1968), reproduces the wrappers of the parts; and see the note preceding the reprint of W. J. Hooker and G. A. Walker Arnott, *Botany of Captain Beechey's Voyage* (Weinheim 1965).

The bibliographer's concern with format, variant title-pages and so on may seem to the historian close to antiquarianism; but books are physical objects, and the format in which they appeared can tell us something about the public to which they were addressed—the historian should always try to see original editions of the works he uses. Bibliographers may be able to give us clues to the size of editions of a book, and thus to its circulation, in those many cases where we have no direct evidence; and the variant title-pages may warn us about the practice of reissuing a book with only the title-page changed, thus giving it in a sense a specious date. Claims made on title-pages that this is a revised edition should not always be taken at face-value, for often such things are reprints of more or less accuracy.[1] Title-pages can even be thoroughly misleading about contents as well as dates, as in Robert Hooke's *Posthumous Works* which is there alleged to contain his 'Cutlerian Lectures', though in fact it does not. Reprint publishers in our own day have sometimes increased this confusion, by omitting original title-pages or by describing as a second edition what is in fact a facsimile reprint of the first with a brief introduction. When there has been a second edition, a facsimile of the first may even be described as a third edition. The historian blesses those reprinters who clearly state the whereabouts of the book of which they are publishing a facsimile, and who reproduce the original title-page.

There are a few general bibliographies of scientific books;[2]

[1] This is true for example for J. Graunt, *Observations . . . upon the Bills of Mortality*, 5th ed., 1676; R. Hooke, *Posthumous Works*, ed. R. Waller (1705, reprint London 1971)—the reprint, a facsimile of the Royal Institution's copy of the first edition, is described as a second edition; the specious 3rd ed. is T. E. Bowditch, *Mission from Cape Coast Castle to Ashantee* (1819, reprint London 1966).

[2] J. L. Thornton and R. I. J. Tully, *Scientific Books, Libraries, and Collectors*, 3rd ed. (London 1971); my own *Natural Science Books in English, 1600–1900* (London 1972), contains essays and book-lists; the standard work is J. C. Poggendorff, *Biographisch-literarisches Handwörterbuch zur Geschichte der exacten Wissenschaften*, 7 vols. in 14 (Leipzig 1863–1961). For great books, see B. Dibner, *Heralds of Science* (1955, reprint Cambridge, Mass. 1969), for current books,

and others dealing with technology and with aspects of natural history, such as bird books and flower books. The distinction between the book-collector and the natural historian was not easy to draw in the eighteenth and nineteenth centuries; some works with beautiful illustrations are and were of small scientific importance, while others contain first descriptions and illustrations of type-specimens from which species were first described, and therefore have a scientific as well as an aesthetic value. The bird-books of J. J. Audubon and of John Gould fall into this second category; while Thomas Bewick's *Birds* and *Quadrupeds*, delightful as they are and widely used by amateurs as they were, fall into the first. Bibliographies are often written for collectors and therefore emphasise beauty, which may or may not have struck contemporaries, and rarity, which may be an indication that few contemporaries saw or knew of the book; but they do provide a very useful entry into a field. Once a field has been entered, bibliographies of individual authors will be very valuable aids to the historian, relating a man's books to one another and also to papers and manuscripts. There exist bibliographies for many of the greatest scientists,[1] from Hooke, Boyle and Newton at the

national bibliographies such as the *British National Bibliography* and the American *Cumulative Book Index*; E. S. Ferguson, *A Bibliography of the History of Technology* (Cambridge, Mass. 1968); G. Ottley, *A Bibliography of British Railway History* (London 1966).

---

[1] For example, *R. Hooke*, by G. Keynes (Oxford 1960); *R. Boyle*, by J. F. Fulton, 2nd ed. (Oxford, 1961); *J. Evelyn*, by G. Keynes, 2nd ed. (Oxford 1968); *J. Ray*, by J. Keynes (London 1951); *R. Bentley*, by A. T. Barthomew (Cambridge 1908); *Newton*, by J. G. Gray, 2nd ed. (Cambridge 1907); *J. Priestley*, by R. E. Cook (London 1966); *C. Linnaeus*, by B. B. Woodward, W. R. Wilson and B. H. Soulsby, 2nd ed. (London 1933); *G. White*, by E. A. Martin, 2nd ed. (London 1934); *A. L. Lavoisier*, by D. I. Duveen and H. S. Klickstein (London 1954), *supplement* 1965; *J. Smeaton* (engineering designs), by H. W. Dickinson and A. A. Gomme (London 1950); *J. Dalton*, by A. L. Smyth (Manchester 1966); *H. Davy*, by J. Z. Fullmer (Cambridge, Mass. 1969); *M. Faraday*, by A. E. Jeffreys (London 1960); *C. Darwin*, by R. B. Freeman (London 1965); *P. H. Gosse*, by R. Lister (Cambridge 1952); *J. Gould*, by R. B. Sharpe (London 1893); *O. Lodge*, ed. T. Besterman (London 1935).

beginning of our period to Darwin and Oliver Lodge at the end of it; these cast light also upon contemporaries and list lectures founded by or in memory of the great man. Bibliographies should also refer us to works by others in which the subject's doctrines are propagated or attacked; but of course it is too much to expect them to do this except for works devoted explicitly and entirely to such ends. Subject bibliographies, guides to the literature of a particular science, have usually been drawn up with students of that science in mind;[1] the older literature may therefore be cited for its curiosity, and the more recent for its contribution to the modern view, rather than for such contemporary importance as it may have had.

Bibliographies thus tend to list works which are outstanding, or anyway were written by very eminent people; which is not always what we want in our attempts to assess the norm, or even to see what the great men were sometimes driving at, or appeared to be driving at. Here libraries and library catalogues may be very useful indeed. Some of these are modern collections of the writings of and about one man, or books relating to one science; catalogues of these collections really count as bibliographies as described above.[2] Particularly interesting for establishing norms are libraries which were collected in the past and have remained more or less unchanged since. The prime example of this is probably the Pepys Library at Madgalene College, Cambridge.

[1] L. Agassiz, *Bibliographia Zoologiae et Geologiae*, 4 vols. (London 1848–54); H. C. Bolton, *A Catalogue of Scientific and Technical Periodicals, 1665–1882* (Washington D.C. 1885), and *A Select Bibliography of Chemistry*, 4 pts (Washington D.C. 1893–1904); P. F. Mottelay, *Bibliographical History of Electricity and Magnetism* (London 1922); G. E. Fussell, *Old English Farming Books, 1523–1793*, 2 vols. (London 1947–50).

[2] I. Macphail, *Alchemy and the Occult; A Catalogue of Books and Manuscripts from the Paul and Mary Mellon Collection*, 2 vols. (New Haven 1969); J. Ferguson, *Bibliotheca Chemica*, 2 vols. (Glasgow 1906); R. B. Webber and H. P. Macomber, *A Descriptive Catalogue of the Grace K. Babson Collection of the Works of Sir Isaac Newton*, 2 vols. (New York 1950–55); A. N. L. Munby, *Cambridge College Libraries* (Cambridge 1962); P. Morgan, *Oxford Libraries outside the Bodleian* (Oxford 1972).

The vast majority of the books Pepys bought new and had bound; they are still arranged as he had them at the end of his life. As is well known, he used the size of the book as the main factor in deciding where it should go; and even books of the same size, like the two volumes of Boyle on air, may be widely separated from one another. He had a catalogue of subjects as well as of authors, so we can see what he classified as 'Arts and Sciences'; while he had some nice scientific books, it is evident from his library that he was elected President of the Royal Society because he was a powerful man with scientific interests rather than a man of science.

Libraries as unchanged as Pepys' are rare; but there is in Durham, for example, the collection known as the Bamburgh Library which was assembled by a prominent clerical family, the Sharps, in the seventeenth and early eighteenth centuries. This contains about as much science as Pepys' Library; because it has been moved about the classification is not exactly as it was, but again size played a considerable part in it, and such subject classification as there is is very different from a modern one. These libraries indicate how difficult it is to distinguish what we call 'science' from other activities when we go back in time. These collections have not been significantly added to or subtracted from; similar libraries can be found in cathedrals, where they were collected by deans and chapters, and within the libraries of colleges and universities. They are particularly valuable when one can browse among the books rather than order those that one wants, because it is in looking along the shelves that one finds the books that one did now know about already. It is from the collection as a whole that some kind of norm can be determined, and not from some books in it; though it may be interesting, for example, to know that Pepys and Sharp did acquire Hooke's *Micrographia*, and that the dean and chapter of Durham bought some Cuvier.

Most libraries have a history of acquisition, disposal and recataloguing which makes it more difficult to know what the collection looked like at an earlier period; in these cases there may be a printed or MS catalogue which enables us to recover the past content and arrangement of the library.[1] Thus at the Royal Institution in London there are printed catalogues prepared at various times in the nineteenth century, so that we can without great difficulty tell what books were available at a given date; the same is true of other libraries of learned societies, and of provincial institutions. Such libraries often have the advantage over larger collections that the books are on open shelves; there is no doubt that the putting of books in stacks where the reader is not allowed to go is a considerable obstacle to historical research. Great works can always be summoned up; but the historian in an open library of primary sources can expect happy accidents and discoveries which may tell him more about the actual state of science than the great works will.

In a scientific institution, one can hope to illuminate the work done by consulting the books used there; though naturally a man may not have used the books which were in his own library or in the library attached to his laboratory or college. For evidence that somebody owned a book, we may as with Pepys have a diary to prove it, or an extant library which may be kept as a collection or may be distributed within another library, perhaps at his place of work. If this collection was sold, there may be a sale catalogue;[2] all or part of his library may be in the hands of his descendants. A useful guide is the table of subscribers which appears at the front or back of those numerous works of the seventeenth and eighteenth centuries which were published by

[1] There are printed catalogues of Edinburgh University Library; and of the libraries of the Science Museum, and the British Museum (Natural History), at South Kensington; and of the London Library; for example. See also A. Franklin, *Les Anciennes Bibliothèques de Paris*, 3 vols. (Paris 1867–73, reprint Amsterdam, 1968). In general, see R. L. Collison, *Bibliographies*, 3rd ed. (London 1968); T. Besterman, *A World Bibliography of Bibliographies*, 5 vols., 4th ed. (Lausanne 1965–66).

[2] A. N. L. Munby (ed.), *Sale Catalogues of libraries of eminent persons* (London 1971–); P. J. Wallis, of the University of Newcastle-upon-Tyne, is working on subscription-lists of scientific books: see *The British Journal for the History of Science*, 6 (1972), 227–8.

subscription. Not only does this tell us whether a man bought a book—it cannot tell us that he did not, for the subscribers were not the only purchasers—but it enables us to assess the public to which the book was directed. The English edition of Glauber's *Works* (1689), for example, shows a surprising interest among medical men in a book so chemical (and often alchemical); and the various *Natural Histories* of counties show a mixture of local gentry and clergy and of those whom one recognises as men of science. Rather similarly, reports of expeditions sometimes contain lists of those who supported them;[1] thus George Back's account of his Arctic expedition in search of John Ross has such a list, and so does F. L. McClintock's story of his search for Sir John Franklin; we can see to what extent these were supported by scientists, by men in public life, and so on. If we find men of science well-represented, this is an indication that such expeditions were seen as an integral part of science. We may also find, as we do in McClintock's book and in McClure's account of his making a North-West passage, memorials to government urging more action in the search for Franklin which were signed by scientists and eminent persons in other fields and enable us to assess again whether such exploration was seen as science. A posthumous book, like Bishop Heber's *Journal*, may contain a list of those who have subscribed to a monument in his honour; if one were to find this for a man of science it would give some clue to his reputation and to the circles in which he moved.

But books are chiefly important for their actual content, rather than their subscription-lists; and it is to their content that we should now turn our attention. Bibliographies in modern books—secondary sources or new editions of primary sources—

[1] G. Back, *The Arctic Land Expedition* (London 1836); F. L. McClintock, *The Fate of Franklin* (London 1859); R. McClure, *The North-west Passage*, ed. S. Osborn (London 1856); R. Heber, *Narrative of a Journey through the Upper Provinces of India*, 2 vols. (London 1828). For other information on books as physical objects, see e.g., P. Simpson, *Proof-reading in the 16th, 17th and 18th centuries* (Oxford 1935); R. McLean, *Victorian Book Design*, 2nd ed. (London 1972).

will help us to enter a field, and so will the collections in small open libraries; but after that it will be chiefly a matter of following up references in the primary books. It should not be supposed that scientists of the seventeenth and eighteenth centuries, or indeed of later times, always paid their intellectual debts by giving a proper reference to works they had used. Laplace, for example, did not cite some of the most valuable writings of his predecessors in his own work on the stability of the solar system, as his translator, the American sea-captain Nathaniel Bowditch, pointed out.[1] Bibliographies have often been more or less copied from another work, and that a work is referred to is no guarantee that it has been actually read; anyone who has done any work in the history of science will be familiar with erroneous references which keep coming up. That somebody copied an erroneous reference from a book or paper is of course excellent evidence that he consulted the source of the error and presumably indicates that the author erroneously cited was felt to be a great man with whose work one ought to be familiar. In the earlier part of our period, the references will almost always be inadequate by modern standards, and even in the nineteenth century references in books are often excessively brief and incomplete, although those in the best journals are adequate. The historian becomes familiar with the mere mention of a name, or maybe references to 'Smith's *Travels*, 8° edition', with no author's initial, date of publication, or page number; and he knows that if he were to find the book the chance that it has an index is very small. It is here that the catalogues of large libraries become valuable.

It may be that there is a subject bibliography which will enable one to find the right 'Smith', and with luck even to work out what must be meant by 'the octavo edition', and perhaps where one is to be found.[2] Or it may be that one can find a 'Smith' of

[1] N. Reingold, *Science in 19th-century America* (London 1966), 11ff.

[2] e.g. *Travel . . . from the Library of Major J. R. Abbey*, 2 vols. (London 1956–57). *Dictionary of National Biography* is wrong, for instance, on the initials of Captain J. H. Tuckey, R.N.; R. L. E. Collison, *Encyclopedias: their history throughout the Ages, a bibliographical guide* (New York 1964); S. P. Walsh,

appropriate date who was a traveller in a biographical dictionary, such as the *Dictionary of National Biography*; this last is usually a fairly reliable guide to initials, publications, and so on, although it would be unwise to rely upon it implicitly. It is all too likely, however, that this 'Smith' will be too obscure to have appeared in *Dictionary of National Biography*, and that his works will not be found in a modern subject bibliography. It is perhaps worth looking in some encyclopedias contemporary with the reference; with luck our Smith may be referred to in the *Encyclopedia Britannica*; the *Metropolitana* or the *Edinburgh Encyclopedia*, though it will probably be *en passant* and he may well be called 'Mr Smith', so that we shall not find out his initials. We must therefore turn to library catalogues; beginning with smaller ones, such as that of the London Library or of Edinburgh University Library, but if necessary going through the entries in the *British Museum Catalogue* or the *American Union Catalog*, or that of a copyright library.

For a name like 'Smith', this is something undertaken as a last resort; with an uncommon name, especially if we know the man's approximate dates and have an idea of his interests, it need not take much time. A man's interests may have been very diverse, and may even seem to us incompatible. He may, like Thomas Beddoes, have written a best-selling moral tale as well as medical and chemical works; or like James Parkinson he may be famous for his clinical studies and his palaeontological writings; so one must not skip too hastily down the columns in the catalogues. Hunts such as these often bring to light interesting and unexpected things; but one owes it to one's successors as far as possible not to make them wade across the same ground, and therefore to make one's own citations sufficiently full. For this, place and date of publication, and the name and initials of the author, are essential; a short title can usually be given for those

---

Anglo-American General Encyclopedias, 1703–1967 (New York 1968); and see A. N. L. Munby, *History and Bibliography of Science in England: the first phase, 1833–45* (Berkeley and Los Angeles 1968).

works whose full title is several lines long, but a bibliography or library catalogue should be consulted to make sure that the short title is unambiguous, for authors are sometimes given to using similar titles for various books. Sometimes there is some confusion as to who exactly was the author; thus what is usually referred to as Anson's *Voyage* was composed, according to its title page, by the chaplain, Richard Walter, but in fact seems to be largely by Benjamin Robins.[1] William Combe similarly acted as ghost-writer for scientists and others; but this need not be mentioned in a citation, which is only to help readers find the book referred to. Where an edition other than the first is referred to, this should be made clear; specious modernity is otherwise given to nineteenth-century authors reprinted in the twentieth century, for example, and the reader who happens to know that the author referred to had been dead for fifty years when the book is alleged to have appeared will suspect that the historian has not done his job properly. If a book or edition is excessively rare, it is very useful to say where it was found.

We might have been pursuing Smith's *Travels* because of an interest in technology, anthropology, navigation, or the relation of science to government at some date within our period; and the time has now come to draw up certain categories of books which are valuable in their different ways as sources for the historian of science. Any classification, it must be said at the outset, is bound to be artificial; to separate textbooks from monographs or popularisations, for example, is in many cases easy but often has to be arbitrary. In determining a norm, that is, in discovering what was taken for granted and which scientists enjoyed most prestige, textbooks are a valuable source indeed. We tend to think of science as a matter of voyaging through strange seas of thought—that is, of hypotheses, theories and daring examples of inductive and deductive logic. This is indeed a part of the story;

[1] G. Williams (ed.), *Documents relating to Anson's Voyage round the World, 1740–4* (London 1967); L. Heaps (ed.), *Log of the Centurion: Based on the papers of Captain Philip Saumarez* (London 1973). On Combe, see H. W. Hamilton, *Dr. Syntax* (London 1969).

but much science is a matter of the arranging of well-established facts, and it is this part of the subject which is the province of the textbook. If the scientific journal with its brief articles brought about 'normal science' with its accumulation of data, the scientific textbook is the vehicle by which this science is conveyed to the rising generation.[1] Occasionally textbooks are original in approach; Lavoisier, as we have remarked, introduced his reform of chemistry to the world in a textbook. But as a rule they are and always have been rather dull compilations, differing from one another in the up-to-dateness of their information, the clarity with which it is organised, and the knowledge the author takes for granted; and always some distance behind the frontier of knowledge, which by the time it gets into the textbooks has become assimilated into the body of accepted science.

This is perhaps a caricature, but it is probably true that different textbooks in the sciences do and did differ from one another much less than do the books that, at a comparable stage, those studying history or literature would use. A feature of textbooks, and of new editions of textbooks, is that they make their predecessors obsolete; older textbooks contain poorer values for the atomic weights or melting points of substances than do new ones, they contain generalisations which can no longer be accepted, and they use the language of theories no longer in vogue. To look at successive editions of a textbook which was for many years standard, such as Thomas Thomson's *System of Chemistry*, is to see the progress of the science displayed; and the best evidence of activity in a science is rapid change in its textbooks. In works of this kind we can hope to see how fast new discoveries were incorporated into the science, and how new insights led to new arrangements of the materials. Thus in chemical textbooks between about 1800 and 1850 we can see how the theory that heat was an effect produced by the weightless fluid 'caloric' gave way to the view that it was the motion of particles which

[1] T. S. Kuhn, *The Structure of Scientific Revolutions*, 2nd ed. (Chicago 1970); these points are briefly and amusingly made in Kuhn's paper in A. C. Crombie (ed.), *Scientific Change* (London 1963), 347–95.

was responsible;[1] and we can see how chemists came increasingly to use the atomic theory, although they consistently urged that it should only be fully accepted 'in a form divested of all hypothesis'. We also find that the great series of recipes which had about 1800 constituted organic chemistry, began in later textbooks to be increasingly systematised and ordered, so that by the end of the nineteenth century this branch of the science had become the most orderly part of it. Textbooks can also help to indicate on which side of an academic frontier a given science may be found. Thus we know from correspondence that Joseph Henry's electromagnets were required for classes in chemistry rather than in natural philosophy, and that Faraday was considered a chemist by his contemporaries; and we duly find in textbooks of the first half of the nineteenth century that electricity is treated in works on chemistry.

Part of the reason for this seems to have been that chemistry had by the late eighteenth century emerged as a fairly well-defined science, with a great number of facts to be organised; the time was therefore ripe for textbooks of chemistry. Chemistry also promised to be useful, in industry, in agriculture, in mineralogy and geology, and later in the analysis of food, drugs and water; and there was therefore need for textbooks at various levels to supply those who had to learn the subject. Natural philosophy on the other hand was more mathematical and seems to have had fewer textbooks until the mid-century when the principle of conservation of energy led to the emergence of the science we call physics, in which natural philosophy was united with other sciences such as optics and electricity which had earlier seemed distinct from it. At all events, the series of textbooks at different levels from the elementary to the advanced, with their small changes in successive editions appearing every few years, are

[1] R. Fox, *The Caloric Theory of Gases* (Oxford 1971); R. J. Morris, 'Lavoisier and the Caloric Theory', *British Journal for the History of Science*, *6* (1972), 1–38; S. G. Brush, 'The Wave Theory of heat', *British Journal for the History of Science*, *5* (1970), 145–67. On the atomic theory in textbooks, see W. H. Brock (ed.), *The Atomic Debates* (Leicester 1967), essay I.

a prominent feature of the sciences since 1800 or so; the historian of an earlier period will meet popularisations, but he will encounter few textbooks. Science was learnt from popular or original writings, without the need for textbooks to fill the gap between them.

Allied to the textbook is the technical manual. In any science of experiment or observation, it is clearly vital to know how to use the instruments and apparatus at one's disposal; and the historian cannot as a rule come to grips with past science until he has a good grasp of the experimental methods available to those whose work he is studying. In the seventeenth and eighteenth centuries there were available, for surveyors, navigators and astronomers, treatises on the use of instruments and globes; later came descriptions and instructions for the use of chemical apparatus. We shall discuss these works further when in a later chapter we discuss surviving apparatus as a source for the historian; in the present context such books are valuable as indicating the kind of experiments carried on, the accuracy with which different processes could be performed, and the meaning of technical terms such as 'lixiviation' which have passed out of ordinary usage. A good example of a manual of this kind is Faraday's *Chemical Manipulation*,[1] (1827), which was not designed to describe any particular chemical reactions but to indicate how chemical processes could be best carried out. Books of this kind can tell us what kind of investigations were fashionable, and therefore cast light on the state of theory at the time. Thus Lavoisier in his *Elements of Chemistry*, (1790) had described chemical manipulation, because he believed that if his readers carried out his experiments exactly as he had, they would come to his conclusions. He gave instructions for performing experiments in which the heat produced in chemical reactions would be measured; but his successors found that these,

[1] See note I, p. 133, for bibliographies of Lavoisier and Faraday. A useful introduction to navigational instruments is the Greenwich Maritime Museum's guide, *Man is not Lost* (London 1968). On electricity and chemistry, see T. H. Levere, *Affinity and Matter* (Oxford 1971); and C. A. Russell's introduction to the reprint of J. J. Berzelius, *Essai . . . 1819* (New York 1972).

which depended upon seeing how much ice was melted when a given weight of various substances reacted together, did not yield constant and repeatable results. This approach to chemical energetics was therefore for a time given up, and Faraday a generation later did not describe experiments of this kind. But whereas Lavoisier had been little concerned with electricity and did not deal in his *Elements* with electrical experiments, Faraday devoted considerable space to it; for Volta, Davy and Berzelius had shown that electrical and chemical changes were indissolubly connected.

By the mid-nineteenth century one finds again an interest in heat as an agent in, and a measure of, chemical change; as we can see from the English version of Leopold Gmelin's *Handbook of Chemistry*, which appeared between 1848 and 1852. This is a work which is difficult to classify, and which is as invaluable for the historian as it was (and later editions still are) for the chemist. It comes somewhere between a textbook and a laboratory manual; for it would be impossible to read through such a compendium of information, and yet it contains bibliographies and theoretical discussions as well as the latest values of constants such as atomic weights and specific heats. If we want to know what was known about the chemistry of platinum, or whose measurement of the vapour-pressure of turpentine or carbon disulphide was believed to be the best available, we turn to Gmelin's *Handbook*.[1] Textbooks in general are a useful source of names to pursue further; Gmelin with his bibliographies and tables is exceptionally valuable in this respect, for in his work we meet those ordinary scientists who enjoyed prestige in their own day but who are now forgotten because their measurements have been superseded, and they did not in any way revolutionise the science in which they laboured. Gmelin's *Handbook* was published by the Cavendish Society, which produced books rather than a journal; similarly the Ray Society published works on the biological sciences, including

[1] L. Gmelin, *Handbook of Chemistry*, tr. H. Watts, 6 vols. (London 1848–52). The Cavendish Society, *Works*, 30 vols. (London 1848–72); R. Curle, *The Ray Society: a bibliographical history* (London 1954).

Lorenz Oken's speculative *Physiophilosophy* in 1843 and Darwin's standard work on barnacles in 1851–4. Publications of these societies had a certain authority and must have recommended themselves to the eminent men on their councils; and as they went to the members one could find out who got them. It seems fair to say that their publications were prestigious and became well known among all those doing serious work in chemistry or biology.

Gmelin's translator was Henry Watts, who later himself compiled a *Dictionary of Chemistry*.[1] Scientific dictionaries form a useful genre, going back certainly to the eighteenth century; they differ from a work such as Gmelin's in that they are arranged in an alphabetical rather than a logical manner. This makes it easier to look up a term such as 'heat' or 'atom' to see what sense it conveyed at the date when the dictionary was compiled—or usually rather before, since lexicographers tend to be conservative. We can sometimes find out from dictionaries about technical terms and processes employed in industry; this is particularly true of the chemical dictionaries. We are told that it was from William Nicholson's *Dictionary of Chemistry* (1795) that Humphry Davy began to teach himself chemistry; this may be true, but in general these dictionaries are forbidding compilations, in the best of which one might browse but from which one would not expect to learn a subject; they were works of reference. A very useful one is Charles Hutton's *Mathematical and Philosophical Dictionary* (1795–96), which has the bibliographical peculiarity that the second volume appeared before the first; this has lengthy articles on theoretical terms and is a valuable guide to the science of the 1790s, giving arguments on both sides of disputed questions. For the whole range of sciences at a slightly later period, one may with great profit consult Thomas Young's *Lectures on Natural Philosophy*,

[1] H. Watts, *A Dictionary of Chemistry*, 5 vols. (1870); D. Layton, 'Diction and Dictionaries in the Diffusion of Scientific Knowledge', *British Journal for the History of Science*, 2 (1965), 221–34. T. Young, *A Course of Lectures on Natural Philosophy and the Mechanical Arts*, 2 vols. (London 1807, reprinted New York 1971).

an extremely erudite course which passed above the heads of those to whom it was directed at the Royal Institution, as indeed it would have passed above the heads of any audience. The lectures themselves are full of interest, when read; but to the historian it is the 'Systematic Catalogue of Works relating to natural philosophy' in the second volume which will be of most value as a collection of source materials compiled by one of the ablest and best-read men of science of the early nineteenth century.

Published lectures were during our period an important source; during the eighteenth and nineteenth centuries, especially in Britain and America, much of the dissemination of science was done through lectures, and the best of these, such as J. T. Desaguliers', make excellent reading and give us a good idea of how science was presented.[1] Where we do not find a full publication of courses of lectures, we may find a published syllabus, sometimes illustrated to show (and perhaps to advertise) the apparatus employed; this can be useful, for it shows what the lecturer considered important and attractive and allows us to estimate how much time he devoted to various topics. Academics in eighteenth-century England do not seem to have delivered many courses of lectures on science, and publishable courses were few and far between; but we do have book-lists from eighteenth-century Cambridge in Christopher Wordsworth's *Scholae Academicae* (1877). In Holland and in Scotland the formal teaching of science —particularly medicine, to which chemistry, botany and zoology

[1] A. Thackray, *Atoms and Powers* (Cambridge, Mass. 1970), chap. VIII; W. D. Miles, 'Public Lectures on Chemistry in the United States', *Ambix*, *15* (1968), 129–53; J. T. Desaguliers, *A Course of Experimental Philosophy*, 2 vols. (1734–44); a pirated *System* had appeared in 1719. F. Hauksbee (the younger), *A Course of mechanical . . . experiments* (London *c.* 1730), advertised a course; Thomas Garnett and Davy published syllabuses of their courses at the Royal Institution. G. A. Lindeboom, *Herman Boerhaave* (London 1968); J. B. Morrell, 'The University of Edinburgh in the late 18th century', *Isis*, *62* (1971), 158–71, and 'Thomas Thomson', *British Journal for the History of Science*, *4* (1969), 245–265. J. Black, *Lectures*, ed. J. Robison (Edinburgh 1803); *Notes*, by T. Cochrane, ed. D. McKie (Wilmslow 1966). J. Walker, *Lectures on Geology*, ed. H. W. Scott (Chicago 1966).

were ancillary—was more advanced, and we find some notable courses of lectures delivered at Leyden or Edinburgh being published. These are and were particularly valuable when the author was reluctant to go into print, so that his works might never have appeared had it not been for the threat, or the actual publication, of a pirated version compiled from the notes of a student. Desaguliers' lectures had been thus pirated, and so had Herman Boerhaave's which appeared in English translation in 1727; both these authors then produced authorised versions of their courses, although in Boerhaave's case the pirated version, of which a second edition appeared in 1741, seems to have been the one most used by contemporaries. Sometimes, as with the lectures of Joseph Black who taught chemistry at Glasgow and then at Edinburgh, the edition is posthumous; this raises difficulties if the manuscript has disappeared because we do not know how much the editor rearranged or amended what was presumably a collection of notes revised each year or at least from time to time; the published version may never have been delivered as it stands, or at any rate not as one course. Sometimes manuscript notes, made either by the lecturer or by one of his audience, come to light; and an edition of these can be very valuable indeed because —especially if it is the notes of a student—we can know that this was the course as delivered in a particular year. Students vary in the accuracy with which they takes notes, and the transcript must therefore naturally be used with care; we must not put too much weight upon it, but where it agrees with what the lecturer has written at another date we can probably rely on it. A student's notes from Black's chemistry lectures of 1767–68 have recently been published, with useful annotations; as have the geology lectures of his Edinburgh contemporary John Walker, of which several manuscripts survive in various hands.

Lectures, textbooks, and manuals indicate how sciences were taught and what was the state of the various sciences at their date, though they may only indicate a best practice rather than a general practice. Thus Robert Hooke in his *Micrographia* of 1665 had described the form and use of a microscope, and microscopes were used by Nehemiah Grew, by Marcello Malpighi, and by Antony van Leeuwenhoek in researches at the end of the seventeenth century. In 1753 H. Baker's *Employment for the Microscope* was published; and yet when in 1816 Captain J. H. Tuckey's expedition was sent out to the Congo, none of the natural historians appointed through the Royal Society could 'either describe or draw' marine creatures from the microscope.[1] It was necessary for M. J. Schleiden thirty years later to remind his readers that the microscope was as necessary to biology as the telescope to astronomy, and to urge them furthermore to use a compound microscope rather than a single lens. We may also find that there was no uniform practice—that the various standard textbooks differ from one another. In chemistry in the 1840s and 1850s for example, there was no agreement about notation and about the atomic weights of the various elements. Chemists could agree within limits about analyses but not about the formulae to be used; water, for instance, contained eight parts by weight of oxygen to one of hydrogen, but its formula could be written HO or H_2O, the choice depending upon various analogical arguments. If the formulae of reactants and products were thus uncertain (and chemists even differed as to whether the symbols represented particles, weights or volumes), attempts to find and classify reaction mechanisms naturally enough led further and further into uncertainty.

The historian's task here is to account for this state of affairs; and this will lead him away from textbooks and towards publications of more originality, and perhaps general interest. For while textbooks replace one another, and it is a test of the importance of a textbook to see how many of its predecessors it killed off, there are some monographs in the sciences which live on. But it

[1] S. Bradbury and G. L. E. Turner, (ed.) *Historical Aspects of Microscopy* (Cambridge 1967); J. K. [*sic*] Tuckey, *An Expedition to explore the River Zaire* (London 1816), 49; J. M. [*sic*] Schleiden, *Principles of Scientific Botany*, tr. E. Lankester (London 1849), 575–91; on chemical formulae, see M. P. Crosland, *Historical Studies in the Language of Chemistry* (London 1962); C. A. Russell, *The History of Valency* (Leicester 1971); and my *Atoms and Elements*, 2nd ed. (London 1970).

should be pointed out that despite the efforts made to persuade scientists to read classics from the history of their science, many never do; the history of science is littered with great names whose works are very rarely looked at. This is a pity because many students of science do find it exciting and entertaining to look at the primary documents rather than to be content with a brief historical account from a modern textbook of the science. To concern oneself at all with the history of science and not to work with primary sources is like trying to learn an empirical science without doing any laboratory work; one can get only the shadow and not the substance.

To help those whose time and access to libraries is limited, there are source books of various kinds now available;[1] these are most valuable when they print extracts of reasonable length. To get a single paragraph, often in modern translation, is not a good way of finding out what a man thought and why; and ideally one should have complete papers or chapters from books. There are two kinds of sourcebooks; those which cover a lot of development in a single field, from which one may hope to see the dialectical manner in which science progresses; and those which collect extracts of great importance from across the spectrum of the sciences, so that there will be something to interest everybody although there will be little or no relation between the various excerpts. One would hope that reading brief extracts from books, or single papers, would be a process sufficiently tantalising to send readers to the complete texts or to other papers; but this is often merely a pious hope, and of course it may well be that scientists in the past did not read books

[1] The series of *Sourcebooks* includes *Physics* (New York 1935), ed. W. F. Magie; *Chemistry* (New York 1952), ed. H. M. Leicester and H. S. Klickstein; *Greek Science* (Cambridge, Mass. 1958), ed. M. R. Cohen and I. E. Drabkin. Collections on a theme include M. P. Crosland, *Science of Matter*, and L. P. King, *History of Medicine* (Harmondsworth 1971); and the *Classical Scientific Papers* series of longer pieces, *Physics*, ed. S. Wright (London 1964), and *Chemistry*, 2 vols. (London 1968–70), of which I am the editor. A general sourcebook is R. S. Westfall and V. E. Thoren, *Steps in the Scientific Tradition* (New York 1968).

right through but picked the chapters which interested them, although this is not a procedure that one would, except for reference books, recommend.

With works of a forbiddingly formal kind, such skipping about has been recommended on the best authority; thus Newton wrote of his *Principia*[1] to Richard Bentley, who was hoping to use it in a course of lectures against atheism, that 'it's enough if you understand the propositions with some of the demonstrations which are easier than the rest'. Those who used Newton's name and discoveries in their poetry or philosophical writings must usually have found themselves in the same position as Bentley. There are, on the other hand, some great works which are deceptively simple. The prime example is Darwin's *Origin of Species* (1859); and Lyell's *Principles of Geology* (1830–33) might also come into this category. Where authors such as these have transformed the way their successors have looked at the world and have argued their case with considerable forensic skill, it is hard not to see their opponents as stick-in-the-muds or as wilfully obtuse. To read a hostile review of Newton, Darwin or Lyell requires a certain willing suspension of disbelief; but this is an effort which must be made if we are to understand the writings of these authors against the background of their own times. We can then see what questions they were in fact answering and raising; these questions will no doubt be more like those we ask than were those of their predecessors and sceptical contemporaries, but in a scientific advance there is usually some loss as well as gain in generality of explanation; and the questions asked by even the most able and original scientists of the past will not be exactly the same as those we ask, for science is a dynamic

[1] I. B. Cohen (ed.), *Isaac Newton's Papers and Letters on Natural Philosophy* (Cambridge 1958), section IV; this is a valuable sourcebook. M. Nicolson, *Newton demands the Muse* (Princeton 1946); W. P. Jones, *The Rhetoric of Science* (London 1966). L. Agassiz, *Essay on Classification*, ed. E. Lurie (Cambridge, Mass.), reprints a work by one of Darwin's noteworthy opponents; on a Darwinian apostate, see J. W. Gruber, *A Conscience in Conflict; the life of . . . Mivart* (New York 1960).

enterprise and time does not stand still. We must therefore expect to find that even great men made what we would regard as mistakes or hasty judgements, and that some contemporaries will have put a finger on some at least of these, as well no doubt as derided things which to us seem eminently sensible.

The historian must not be content with showing that an eminent scientist neglected evidence he could have taken note of, or that his work was not given due attention. He must try to show why this happened; and this will involve him, for example, in finding out how widely a book circulated: how many (and if possible how large) editions it appeared in; whether it was translated, and if so whether at once or many years later; where and when reviews appeared, and what points were taken up in them; and how soon the book appeared in standard bibliographies or review articles, and under what heading. Thus Lyell's *Principles* and Darwin's *Origin of Species* rapidly went through many editions and could have been ignored by nobody;[1] while Christiaan Huygens' *Treatise on Light* was not translated until two centuries after its author's death and did not—despite Huygens' eminence—play an important role in the history of optics. Davy's *Agricultural Chemistry* (1813), was soon translated into foreign languages; his curious *Consolations in Travel* (1830), was in 1869—forty years after his death—translated into French, when astonishingly enough it seems to have gone through nine editions. Hans Christian Oersted's little book on the unity of force was mauled by Thomas Thomson in his journal *Annals of Philosophy* in 1815, though he admitted he had not seen the original; in 1819 he made amends by a much more sympathetic account, so that Oersted's name was familiar to English-speaking scientists when in 1820

[1] See the bibliographies of Darwin and Davy, note I, p. 133. Editions of Lyell's *Principles* give details of earlier editions; and see L. G. Wilson, *Charles Lyell* (New Haven 1972–). C. Huygens, *Treatise on Light*, tr. S. P. Thompson (London 1912). On Oersted, see B. S. Gower, 'Speculation in Physics', *Studies in the History and Philosophy of Science*, *3* (1973), 301–56; on Huygens' popular writings, my 'Celestial Worlds Discover'd', *Durham University Journal*, *56* (1965), 23–9. R. Olby, *The Origins of Mendelism* (London 1966).

he vindicated his belief in the unity of force with the discovery of electromagnetism. Many reviews of the early nineteenth century consisted of extracts from the book which to our mind were sometimes judiciously chosen and sometimes not, but do give some indication of what contemporaries saw as most interesting or original; it was probably all that many of them ever read. Gregor Mendel's work on heredity, which appeared in a journal rather than a book, was little read and was classified as a contribution to horticulture where it did appear in bibliographies; no doubt books have sometimes been similarly put in the wrong category.

Scientific classics are often very formal in their presentation, and it can be valuable to look at a man's lighter publications in which he is less on his guard. Thus in Huygens' *Cosmotheoros*, or *Celestial Worlds Discover'd*, we find ideas about probability and analogy, and a use of teleological argument, that go well with what is really a work of science-fiction describing the inhabitants of the various heavenly bodies. It is a problem to decide how seriously to take works such as these; but they do sometimes illuminate the most exciting aspect of science, where a creative and disciplined mind comes to grips with matter. For we must never forget that while this is not the whole of science, it is the reason why most people study it; science is a profession with a status and an hierarchy, but it is more than that. In writings like that of Huygens, or like Boyle's on the utility of unsuccessful experiments,[1] at the beginning of our period, and G. H. Hardy's *Mathematician's Apology* at the very end of it, we do feel that we are being taken behind the scenes. A curious example of this genre is Davy's *Consolations in Travel*, which we have already encountered; in this work we see a scientist who was a friend of Coleridge, Wordsworth, Southey and Scott raising in dialogue-form such problems as the immortality of the soul, technological

[1] See note I, p. 133 for Boyle's bibliography; G. H. Hardy, *Mathematician's Apology*, ed. C. P. Snow (Cambridge 1967); see my papers, 'The Scientist as Sage', *Studies in Romanticism*, *6* (1967), 65–88; 'Dynamical Chemistry', *Ambix*, *14* (1967), 179–97; and 'Chemistry, Physiology and Materialism', *Durham University Journal*, 64 (1972), 139–45.

progress, the adequacy of materialism in science, the nature of time, and the kind of explanations needed in geology. Davy, like any other scientist, refuses to fit any of our categories once we look closely at him; and his less formal works encourage us to take this close look and thus to learn more about him and his milieu.

Scientists have often also been their own popularisers, which is a slightly different matter although in popular works men are often prepared to go further than in their formal productions. Men may try to popularise their general approach to nature, or to make known some particular discoveries; often these categories cannot be neatly separated. Boyle's various writings on the corpuscular philosophy might fit into the first category, as does John Herschel's famous *Preliminary Discourse* on scientific method which was one of the great standard works of the nineteenth century.[1] At the end of our period, Einstein wrote a popular work on relativity, though opinions differ as to whether this would enlighten or further confuse the layman. With the rise of mathematics on the one hand and of complicated apparatus and experimental procedures on the other, it became increasingly difficult for the layman to understand what scientists were talking about, and the problem became serious also for men of science working in other fields. Where a man has the ability to write clearly and to express his insights in simple and vivid language, he is likely to be his own best populariser.

Laplace wrote popular accounts of his astronomical work proving the stability of the solar system, and of his studies in statistics.[2] In the former he gave an account also of his nebular

[1] See the introduction by M. Partridge to the reprint of J. F. W. Herschel, *Preliminary Discourse, 1830* (New York 1966); A. Einstein, *Relativity* ... (London 1920).

[2] P. S. Laplace, *Exposition du Système du Monde* (Paris 1796); tr. J. Pond, 2 vols. (London 1809); *Essai Philosophique sur les Probabilitiées* (Paris 1819): tr. (into English) F. W. Truscott and F. L. Emory (New York 1951); see W. Whewell, *Astronomy and General Physics* (London 1833), chap. VI; J. F. W. Herschel, *Outlines of Astronomy* (London 1849). The *Encyclopedia Metropolitana*, 29 vols. (London 1817–45), carried book-length articles on the sciences; and eminent

hypothesis of the evolution of the solar system which was too wild to appear in a great work of applied mathematics; and in the latter he applied statistics to such quantities as the decisions of law courts, in what seems an extraordinarily *a priori* manner for one who held high government office under Napoleon. John Herschel also wrote a popular work on astronomy; and like many of his contemporaries he wrote articles for encyclopedias which were sometimes of book length. Herschel and his younger contemporaries T. H. Huxley and Lord Kelvin also published various courses of popular lectures; and we also find appearing in print series of lectures by different speakers. Lectures or books directed at working men or at a lay audience can in the nineteenth century usually be distinguished without difficulty from those addressed to an assembly of scientists expert in the field; this is often less easy in the first half of our period when a useful indicator is that in popular writings by the 1660s tags or quotations from the learned languages were translated into the vernacular.

Popularisations differ from textbooks in that they would not be aimed to produce experts, though many scientists began their career by reading or even writing popular science. The field is thus very wide; it would include most works of natural theology as well as what we think of as popular science. Natural theology was indeed usually written by those who had done some original scientific work;[1] but the most famous author of all, William Paley, was not a scientist and would from our point of view have to count as a populariser. In the eighteenth century the Newtonian system had attracted a large number of people to serve under its

men wrote too for A. Rees, *Cyclopedia*, 39 vols. (1819–20), and for the various editions of the *Encyclopedia Britannica*.

---

[1] W. H. Brock, 'The Selection of Authors of the Bridgewater Treatises', *Notes and Records of the Royal Society*, *21* (1966), 162–79. On Newtonians, see L. Laudan's introduction to the reprint of C. Maclaurin, *Sir Isaac Newton's Philosophical Discoveries, 1748* (New York 1968). J. Priestley, *The History and Present State of Electricity*, reprint of 3rd ed. (1775), ed. R. E. Schofield (New York 1966); E. C. Patterson, 'Mary Somerville', *British Journal for the History of Science*, *4* (1969), 311–39.

banner, and its popularisers included Voltaire and Count Algarotti as well as numerous lesser lights in England. Sometimes, as with Joseph Priestley, writing what was intended to be a popular account of the history and present state of a science might lead into research and on to fundamental discoveries; but as a rule the popular writer remains on the outskirts of the scientific community. An exception might be Mary Somerville, whose writings on the science of the 1830s and 1840s were greatly respected by scientists in her own day, and who was honoured by scientific societies; she must have succeeded in explaining to men of science what their colleagues were up to, and in getting across a broad view of the sciences in their connexion with one another. At another end of the scale, we find popular works directed at children; these often took the form of a dialogue between teacher and pupil, and the most famous is probably Jane Marcet's *Conversations on Chemistry* (1806), which kindled Faraday's interest in the subject. Little books of this kind serve to show what was taken for granted and what seemed most important in the science of the day; and we should not be distracted by what seems to us quaint or stiff in the dialogue.

Among the most popular sciences there has always been natural history; and here it is often difficult to tell which books were addressed to a popular audience and which to a scientific one. Indeed, before the middle of the nineteenth century such a distinction would probably be meaningless; for natural history was a science chiefly carried on by amateurs. In works of natural history the illustrations are often of great importance; for a good picture can tell us more than several paragraphs of description.[1] The natural history painter has a difficult task, for he has to draw attention to the significant details of the animal or plant—that is, those valuable in classifying it—without distorting the picture into a diagram; and he has to illustrate a typical member of the species rather than a particular individual with its own peculiarities. This the photograph cannot do; photographs must always be of one or more individuals and do not therefore compete

[1] W. Blunt, *The Art of Botanical Illustration* (London 1950).

directly with paintings. The same is true of the interesting nineteenth-century technique of 'nature printing', in which the block was prepared from the actual impression of the specimen. The nature-printed leaf stands out from the page in relief, and for illustrating, for example, ferns the process was very successful; but fragile specimens leave poor impressions, and the technique was never widely applied.

The painting of medical, technological and natural history subjects has both influenced and been subject to the general canons of art at different times; and it has depended upon techniques of collecting and procuring specimens, and of printing the pictures in books. It was not until the nineteenth century was well advanced that the practice of painting birds and even animals from living specimens became at all general. Often skins were sent across the world to a painter who mounted them in what seemed to him an appropriate manner in his studio; and it is no wonder that the resulting pictures have a stiff look about them. The collector tried to bag several, in order to give the painter more than one specimen to work from; and also probably for his supper. Plants were easier to preserve, except in very humid climates. By the 1660s, the woodcut had gone out of fashion for scientific illustration and been replaced by the copperplate, either engraved or etched. The woodcut had had the advantage that it was a relief process, so that prints could be made on the same press as that used for type, and pictures and text could easily be printed on the same page. Copper-plates are made by an intaglio process —that is, the depressions rather than the bumps will print—and therefore must be printed separately from the body of the text. They lent themselves to detail which was hard to achieve in the old woodcuts; both techniques, but particularly the copperplates, present us with the problem that as a rule the artist could not do his own engraving, so that the finished product is only partly his own work. About 1800 came two new processes, wood engraving, on the end-grain of boxwood, which particularly in the hands of Thomas Bewick yielded very fine results, and lithography in which prints could be made from a drawing made upon a specially prepared stone. This last technique cut out the engraver.[1] With the big editions of nineteenth-century books, wood-engravings and steel-engravings recommended themselves because of their durability. The relation of techniques to the aims of the artists and scientists engaged in natural history and other fields of science where illustration was important calls out for further investigation; and the sources are very abundant.

While looking at the illustrations, we may notice to whom they are dedicated; for in eighteenth-century works the plates are often separately dedicated and provide an indication of patronage. On turning to the text, we may notice whether it describes certain species, irrespective of where they are found; or whether the fauna or flora of a given area is described. In the latter case, we may note the author being chiefly interested in new species, or 'non-descripts'; or else, with greater generality, in all the species so as to compare the distribution of plants or animals in his chosen locality with that in others. Such discussions of natural history are often appended to the reports of expeditions; and in the next chapter we shall make the transition from books explicitly on science to works which bear upon it. We shall look, that is, at books which illuminate indirectly the history of science, especially natural history, technology, scientific education and government policy towards science.

[1] See M. B. Davidson's introduction to *The Original Water-Colour Paintings by J. J. Audubon for the Birds of America*, 2 vols. (New York 1966). On science, technology, and aesthetics, see F. D. Klingender, *Art and the Industrial Revolution*, ed. A. Elton (London 1968).

CHAPTER 6

Non-scientific Books

Local pride, and an interest—economic as well as theoretical—in the plants, animals and minerals to be found in different localities, found an expression at the beginning of our period in accounts of countries or counties in which topography is blended with history civil and natural. Thus in England, William Camden's *Britannia* had appeared in English translation early in the seventeenth century; an expanded version in a new translation was published in 1695, incorporating remarks on the flora of counties by the great naturalist John Ray and on naval dockyards by Samuel Pepys.[1] In 1602 Richard Carew's *Survey of Cornwall* had appeared; with its mixture of local history and genealogy, folklore, and natural history it was the first of a long line of works on counties; going on to the *Victoria* series which began at the end of our period, in the closing years of the last century, and is still in progress. Some of these works were primarily concerned with natural history and declare this interest in their titles; but even these have as a rule been neglected by historians of science who in confining themselves to more formal treatises have disqualified themselves from understanding important aspects of the development of natural history, and thus of biology and geology.

The most famous county histories of the beginning of our period were those of Robert Plot, of the (Old) Ashmolean Museum in Oxford;[2] his *Oxfordshire* and *Staffordshire* provided

[1] W. Camden, *Britannia*, ed. E. Gibson (1695, reprint Newton Abbot 1971); R. Carew, *Survey of Cornwall*, ed. F. E. Halliday (London 1953).

[2] R. T. Gunther, *Early Science in Oxford*, vol. XII (Oxford 1939); R. Plot, *Oxfordshire* (1677), and *Staffordshire* (1686); the 2nd ed. of the former (1705) was reprinted (Chicheley, Bucks. 1972), including the questionnaire. John Aubrey's *Wiltshire*, written at the same period as Plot's works, did not appear until 1847 (reprint Newton Abbot 1969).

models for his successors to imitate. They are large and handsome volumes, full of curious information and very valuable sources for the science of the day. Plot had prepared a questionnaire to be distributed among those who might have access to knowledge of natural history, antiquities or husbandry; and he hoped, as he declared in his dedication to King Charles II, that his volumes might be the first of a series, asking rhetorically 'whether, if *England* and *Wales* were thus surveyed, it would not be both for the Honour and Profit of the Nation?' Not only does Plot's work show how natural history fitted so happily into the other interests of gentlemen, but it contains matters of interest in the internal history of science; most notably perhaps in the geological parts of the *Oxfordshire*, where Plot after weighing the evidence at great length comes at last to the conclusion that 'extraneous fossils' in the shape of shells, owls or human feet are sports of nature and not really the petrified remains of living creatures.

For the mid-eighteenth century, a representative county history would be the *Cornwall* of William Borlase.[1] His work appeared in two substantial volumes, one devoted to the antiquities and the other to the natural history of his county; both volumes are admirable works of eighteenth-century scholarship. Cornwall is rich in minerals and in prehistoric remains, and these particularly Borlase described, within the framework of current ideas. Like Plot, he had a four-element framework for his natural history, describing the air, the water, and the earth of his county; and then he moved on to the vegetable and animal realms, ending with man. Like his contemporaries, he faced the great problem of identifying species described in the texts of antiquity; but this looms much less than it did in writings of the previous century, and Borlase was up to date in his natural history and even in his entertaining passages on the Druid religion which naturally enough he traced back to the

[1] W. Borlase, *Antiquities of Cornwall* (Oxford 1754; reprint of 2nd ed. of 1769, East Ardsley, Yorks. 1973); *Natural History of Cornwall* (Oxford 1758, reprint London 1970); the reprints have useful introductions; *The Islands of Scilly* (Oxford 1756; reprint (reset) Newcastle-upon-Tyne 1966).

Tower of Babel. Anybody interested in natural history and geology in the eighteenth century should read Borlase as a standard source, and should follow up his authorities.

Units smaller than counties had also been described, and even the delightful *History of Myddle* of Richard Gough contains some natural history.[1] Borlase described the Scilly Isles in a little work praised by Dr Johnson; and the best-known small-scale study of natural history is Gilbert White's *Selborne*, which appeared in 1789 and has been constantly reprinted. It is worth using the first edition of this, or the recent facsimile of it, because it contains the *Antiquities* as well as the *Natural History* and thus gives a rather different balance to the whole. White's correspondent Thomas Pennant brought out natural histories of large regions—Britain, the Arctic, and India (which was naturally very sketchy)—which contrast with those of many contemporary zoologists who included numerous exotic birds or animals from various parts of the globe. But exceptions might be James Bolton who in 1788 published three volumes on the fungi which grow around Halifax, and J. E. Smith whose two splendid volumes on the rarer lepidoptera of Georgia appeared in 1797.

By this time a new interest in counties and their natural history had been kindled by the activities of the Board of Agriculture, which had been set up to stimulate landowners to follow the techniques of the improvers. The leading agriculturalists associated with the board were Arthur Young and William Marshall; and an ambitious scheme was launched for describing the agriculture of each county in Britain. The first series of *Reports* appeared mostly in quarto in the 1790s, and there was a later octavo series which came out either side of 1810;[2] many volumes

[1] R. Gough, *Myddle*, ed. W. G. Hoskins (Fontwell 1968); on Pennant, see A. M. Lysaght, *Banks* (London 1971), 435.

[2] For a list of the *Reports*, see my *Natural Science Books* (London 1972), 123–4; Marshall's *Review and Abstract* and some of the county volumes have been recently reprinted, Newton Abbot. See E. J. Russell, *A History of Agricultural Science in Great Britain* (London 1966); and H. R. Fletcher, *The Royal Horticultural Society, 1804–1968* (London 1969).

were also reprinted at various dates, with or without revision. Many of these volumes were written by local improvers, though Young wrote six; they therefore vary considerably in length and depth, but help us to get a splendid view of regional variety. We find plans of labourers' cottages and illustrations of farm machinery as well as descriptions of soils, indigenous plants, methods of cultivation, and weather-reports; and naturally we find full and sometimes fulsome descriptions of the improving landlords' estates. Marshall in 1818 produced a five-volume 'review and abstract' of the *Reports*, which makes splendid reading; he is spirited in his criticisms, and his comparative approach by topics provides a valuable way into the mass of data, some of it dubious, collected in the *Reports*. The Board of Agriculture was closely linked with other bodies such as the Royal Society and the Royal Institution, as well as with the famous agricultural societies at Bath and in Scotland; and its members duly attended such functions as the sheep-shearing at Woburn. Davy delivered courses of lectures on agricultural chemistry for the board in the opening years of the nineteenth century; these were subsequently published and laid the foundations of an applied science which was of great importance.

Reports such as those of Young were indeed chiefly devoted to agriculture, but also discussed in general the economy of the county, in the tradition of Plot and Borlase. Those who had less rural counties to describe often included descriptions of technology. Thus, for example, Young in his descriptions of East Anglian counties had to talk about pumps; and John Holt mentioned early railways in Lancashire. In John Farey's account of Derbyshire, which appeared in three volumes between 1811 and 1817, we find the first publication of William Smith's discovery that rocks may be dated, and strata followed, from the fossil remains found in them. This was to change the whole emphasis of geology from mineralogy to palaeontology, and thus to make the geological remarks of Plot seem curiously old fashioned. But we should not expect in these agricultural *Reports* to find many passages of great originality; instead they are valuable as showing what was going on in agriculture, a science of clear utility and with connexions to botany, zoology, meteorology, geology and chemistry, and one therefore which commanded a great deal of effort and interest around 1800.

Arthur Young had travelled in France and Italy investigating husbandry wherever he went, and had also followed the recommendation of the great Linnaeus that one should be a traveller in one's own country. *Travels* form an important category of books which can be profitably studied by the historian of science; for the traveller, who later became metamorphosed into the explorer, was expected to keep his eyes open and to notice natural history and geology as well as the culture, industry and military strength of the countries which he visited. Distant travels were often supported by governments; thus in the period of friendship between France and Siam, which was concluded by the revolution of 1688 in the latter country, an embassy which included Simon de la Loubère was sent to Thailand.[1] He wrote a remarkably full report on the country, including a study of Siamese mathematics and astronomy for which he had enlisted the aid of Cassini. Similar studies had been made by the Jesuits in China; but diplomats, doctors, and seamen often seem to have been better reporters than missionaries, whose interests were more practical and circumscribed.

With the eighteenth century came specifically scientific expeditions:[2] such as Edmond Halley's voyages to determine magnetic variation in 1698–1700; the journeys of Maupertuis and

[1] S. de la Loubère, *The Kingdom of Siam* (1688, reprint Kuala Lumpur 1969); on travellers in exotic regions, see R. Dawson, *The Chinese Chameleon* (London 1967); F. Wilson, *Muscovy* (London 1970).

[2] E. F. MacPike, *Correspondence and Papers of Edmond Halley* (London 1937); an edition of a journal of his voyage will be published soon by the Hakluyt Society. H. Woolf, *The Transits of Venus* (Princeton 1959); A. Day, *The Admiralty Hydrographic Service* (London 1967); G. S. Ritchie, *The Admiralty Chart* (London 1967); M. Deacon, *Scientists and the Sea* (London 1971); A. Moorehead, *The Fatal Impact* (London 1966); H. E. L. Mellersh, *Fitzroy of the Beagle* (London 1968). See the Hakluyt Society's editions of the voyages of Byron, Cartaret, Robertson (under Wallis) and Cook.

la Condamine to fix the figure of the Earth in the 1730s, which showed that it was flattened at the poles; and the series of expeditions from many nations to observe the transits of Venus in 1761 and 1769. Cook's expedition to Tahiti and on to New Zealand and New South Wales was sent to observe this second transit and had on board an astronomer as well as Joseph Banks the naturalist and future President of the Royal Society; the voyage fits into the history of British penetration of the Pacific, going back to Drake and Anson, but it was also a scientific expedition with definite objectives to fulfil. The French, with Bougainville and la Pérouse, were at the same period adding to geographical knowledge and showing the flag in what might be profitable regions. George Vancouver was sent to the north-west coast of the American continent to supervise the Spanish evacuation of Nootka Sound, where they had endeavoured to displace the English, and also to survey the coastline from California to Alaska.

In the early nineteenth century, the British government supported an extensive programme of naval survey and exploration:[1] after 1815 this used ships and men that would otherwise have been idle; shortened passage times by increasing knowledge of winds, currents and straits; opened up trade with regions such as South America; and furthered the suppression of the slave-trade. In the Canadian Arctic, and later in the Antarctic, less useful voyages were undertaken in the search for the poles and the North-West passage; this seems to be an unusual example of pure science being heavily supported by government, with national prestige and perhaps the need to substantiate territorial claims in what is now the north-west of Canada also involved. These voyages represent the 'big science' of the nineteenth century: a team of scientists, with technologists and technicians (the officers and crew) working together and making observations. The distinction between the scientists and technologists broke down as naval officers and surgeons became increasingly competent

[1] See Day and Ritchie, note 2, p. 165; L. H. Neatby, *Search for Franklin* (London 1970); G. S. Graham, *Great Britain in the Indian Ocean* (Oxford 1967).

as astronomers and natural historians, so that by the middle of the century there was no longer any need for civilian scientists to go. The reports of the voyages were generally written up by the officer who commanded who as a rule wrote very well and unpretentiously; and there were scientific appendices, often published separately, describing the natural history specimens, perhaps the ethnography, and the observations for latitude and longitude made on the expedition. Government support was forthcoming for the publication of these appendices, though not enough to pay the scientists consulted; they often came from that frequently forgotten but very important group who worked in the great museums and whose names in their own day commanded great respect.

Darwin's voyage on the *Beagle* was a scientific voyage, and as well as writing his own account of it—which forms the third volume of the official report—he supervised the publication of the scientific appendices. As his model of a scientific traveller he took Alexander von Humboldt whose *Personal Narrative* of his travels in Central and South America was one of the great books of the early nineteenth century.[1] Humboldt travelled with the support of the French Government, for otherwise he would have been unable to enter the Spanish colonies; but he and his companion Bonpland travelled without official escorts, so that his journeys were not very different from those of a private person. There had been such travellers before, the most famous being Thomas Shaw, whose *Travels in Barbary and the Levant* had been published in 1738 and soon became a classic. One of the great objectives of travellers in these regions and in the Middle East was to identify and correct the geographical observations of the ancients; and the search for the fountains of the Nile and the Mountains of the

[1] A. von Humboldt, *Personal Narrative of Travels . . .*, tr. H. M. Williams, 7 vols. (London 1814–29); tr. T. Ross, 3 vols. (1847). A Moorehead, *The Blue Nile* (London 1962) and *The White Nile* (London 1960); E. W. Bovill, *The Niger Explored* (London 1968); A. H. M. Kirk-Greene (ed.), *Barth's Travels in Nigeria* (London 1962). D. Middleton, (ed.), *Diary of A. J. M. Jephson, 1887–9* (Cambridge 1969).

Moon, for the source of the Congo and the mouth of the Niger, was one of the great epics of the early nineteenth century which yielded botanical, ethnographical and zoological knowledge as well as geographical.

In Asia too there were scientific travellers, perhaps sent by the Society of Dilettanti like Richard Chandler in 1764 to look at ruins,[1] or further East travelling, perhaps in disguise, with the chief object of gaining intelligence on behalf of the East India Company. Intelligent and well-informed men who had administered territories in the East wrote about them; thus we find Stamford Raffles—who became first President of the Zoological Society of London—writing the *History* of Java; William Marsden, who became Secretary to the Admiralty and was a great friend of Banks, describing Sumatra; and Robert Percival writing about Ceylon, which like the other Dutch possessions in the East had come into British hands in the Napoleonic Wars. Science thus benefited from the opening up of what had been jealously kept *terra incognita* both in Spanish America and in Indonesia, through these wars. In India itself, the writings of travellers like Bishop Heber and administrators like Sir John Malcolm, Mountstuart Elphinstone and William Sleeman are valuable sources of science, and were indeed seen as contributions to science.

Diplomatic missions continued to produce their quota of natural history and descriptive science generally; particularly noteworthy examples being the various embassies to China, John Crawfurd's mission to Siam and Vietnam in 1821, and Henry Yule's—as secretary to Arthur Phayre—to Burma in 1855. So did naval surveying voyages; and the colonisation of Australia led to numerous studies of its geography, geology and natural history

[1] R. Chandler, *Asia Minor, 1764–5*, ed. E. Clay (London 1971); S. Raffles, *Java*, 2 vols. (1817); W. Marsden, *Sumatra* (1783, reprint of ed. of 1811, Kuala Lumpur 1966); T. Horsfield, *Researches in Java* (London 1824), the tapir had been long before described by a Chinese expedition, Ma Huan, *Ying-yai Sheng-lan*, 1433, tr. J. V. G. Mills (Cambridge 1970), 101. On India, see Dharampal, *Indian Science and Technology in the 18th century* (Delhi 1971), which reprints contemporary reports. And see above, p. 95.

from the landing of the First fleet onwards.[1] Similarly the United States and Canada were more or less systematically explored, from the first crossing of the continent by Mackenzie and then by Lewis and Clark, through to the massive and handsomely-produced surveys for the transcontinental railways.

Those who travelled in or explored uncivilised regions named the islands, mountains and rivers which they encountered; a study of their naming has been made for Australia and is a guide to patronage or more often to loyalties and friendships; off Australia the French seem to have named geographical features after men of science, while the English applied the names of those in public life or in the navy. Travellers in civilised regions were expected to gather information, by fair means if possible. This might be information on natural history; thus Linnaeus in Holland had the opportunity to study plants from the East Indies in greenhouses,[2] and L'Heritier in England saw and described exotic species in

[1] J. Stockdale (ed.), *The Voyage of Governor Phillip to Botany Bay* (London 1789); J. Hunter, *An Historical Journal, 1787–92*, ed. J. Bach (Sydney 1968); J. White, *Journal of a Voyage to New South Wales*, 1790, ed. A. H. Chisholm (Sydney 1962); G. Caley, *Reflections on the Colony of New South Wales*, ed. J. E. B. Currey (London 1967). A. Mackenzie, *Journals and Letters*, ed. W. Kaye Lamb (Cambridge 1970); and see H. M. Jones, O *Strange New World* (London 1965), 366ff. for the opening of the West. J. A. Ferguson, *Bibliography of Australia*, 7 vols. (Sydney 1941–69); E. H. J. and G. E. E. Feeken, *The Discovery and Exploration of Australia* (Melbourne 1970).

[2] C. Linnaeus, *Hortus Cliffortianus* (Amsterdam 1737); C.-L. L'Heritier, *Sertum Anglicum*, 1788, ed. and tr. G. H. M. Lawrence (Pittsburgh 1963), for works of natural history, see B. H. Soulsby, *Catalogue of the Library of the Linnean Society*, new ed. (London 1925). On industrial espionage, P. Miller, *The Life of the Mind in America* (London 1966), 298; W. O. Henderson (ed.), *Industrial Britain under the Regency* (London 1968); C. van Oeynhausen and H. von Dechen, *Railways in England, 1826–7*, tr. E. A. Forward, ed. C. E. Lee and K. R. Gilbert (Cambridge 1971); E. T. Svedenstierna, *Tour of Great Britain, 1802–3* (Newton Abbot 1973); N. Rosenberg (ed.), *The American System of Manufactures* (Edinburgh, 1969); G. Head, *A Home Tour* (London 1836); R. Southey, *A Tour in Scotland in 1819*, ed. C. H. Herford (1929, reprint Edinburgh 1972); F. Klingender, *Art and the Industrial Revolution*, ed. A. Elton (London 1968); B. Faujas St. Fond, *Travels in England*, 2 vols. (London 1799).

gardens. But it was industrial information which was perhaps most sought after; and in this area we find very valuable official reports compiled by knowledgeable travellers. These are particularly useful because patent specifications of the early nineteenth century are often vague, and because processes were often not patented anyway, being probably kept as far as possible a secret. Carefully written reports by able foreigners can be one of the best sources for the history of technology. We have examples of surveys of railways in England by German engineers, of iron works in Britain in the Napoleonic period, and of American industry in the 1850s by Whitworth and other engineers from Britain; all of which were the results of official travels with the object of improving home industries. There were also travellers interested in technology, such as Robert Southey who made a tour of Scotland with the engineer Thomas Telford, and Sir George Head; their writings tell us less about actual processes and prices, but in some ways more about the Industrial Revolution generally. Technological subjects also attracted a number of artists of ability; the drawings of coal-mines by T. H. Hair, and of the building of railways by John Bourne, for example, tell us a great deal about what was involved. Encyclopedias of the eighteenth and nineteenth centuries are valuable sources of information about technology, usually containing both illustrations and descriptions; Rees' *Cyclopedia* (1819–20) was particularly strong on science and technology, and Cresy's *Encyclopedia of Civil Engineering* (1847) is another useful source.

Machinery and its products were prominently displayed in the Great Exhibition of 1851 and in those which followed it in Britain and other countries, which can be said to have been uniformly less successful. The enormous official *Catalogue* of the Great Exhibition is a fascinating document from which much may be learned, although in it and in other catalogues one finds only the objects described and then often portrayed out of context. More can be learnt from the lectures which were frequently associated with these great international occasions, and from the published *Reports* of the juries of experts who awarded the prizes in the various sections.[1] The juries usually reported upon products rather than on processes, so that one may read their remarks without becoming much the wiser as to how the prize-winning object was made; but one cannot avoid learning some curious social history from seeing what kinds of things were commended, and why. And while the light thus cast upon technological history is tantalisingly flickering, one can get from the *Catalogues, Prospectuses* and *Reports of Juries* some idea of what industries seemed important and progressive at the time. Scientific and precision instruments were among the objects exhibited; the 1876 exhibition at South Kensington was particularly important here, and a useful series of *Conferences* held in conjunction with it was published in 1877, which includes various reflexions by eminent men on the state of science. There had been lectures also associated with the Great Exhibition which were held and subsequently published under the auspices of the Royal Society of Arts; these make splendid reading for the historian.

Such exhibitions depended upon government support; and the Great Exhibition even made a profit which had to be somehow spent. Scientists and technologists associated with the Exhibition, and particularly Lyon Playfair, urged in lectures and committees that this money in particular, and more money in general, should be invested in science, and especially in scientific education and technical training. Money from the Great Exhibition did indeed go into the purchase of land in South Kensington which was used for museums and for colleges to train teachers and scientists, and towards scholarships, from which Ernest Rutherford is so far the outstanding beneficiary. But in Britain government support for scientific education does seem to have come slowly, though this is

[1] *Great Exhibition, Official Catalogue*, 3 vols. and supplement (London 1851); *Reports by the Juries* (London 1852); *Lectures on the Results of the Great Exhibition* (Royal Society of Arts), 2 series, 1852–53. *International Exhibition, 1862, Official Catalogue* (London 1862); *Reports by the Juries*, ed. J. F. Iselin and P. le N. Foster (London 1863). South Kensington Museum, *Conferences held in connection with the special loan collection of scientific apparatus*, 2 vols. (London 1877); C. Babbage, *The Exposition of 1851* (London 1851).

a profitable region for detailed investigations and for international comparisons. The success of the museum at the Jardin des Plantes in Paris in the years after the Revolution, when such notable men as Cuvier and Lamarck were classifying vertebrates and invertebrates there, led to a call for similar facilities at the British Museum,[1] voiced, for example, by Davy, President of the Royal Society in the 1820s, on his deathbed. The volume of work on natural history done at the British Museum did indeed increase during the subsequent years; this rather unglamorous work of classifying species was of great contemporary importance for both biology and palaeontology and ought to be further investigated if we are to get an adequate view of science in the first half of the nineteenth century. At Kew, similarly, botanical specimens were classified and foreign plants cultivated in the gardens and meteorological data collected at the observatory. Fashion even entered in here; for our ancestors' liking for masses of dark vegetation meant that the ferns of Britain were in the 1840s and '50s extraordinarily thoroughly described, illustrated and cultivated.

Similarly the great French institutions of the Ecole des Mines, the Ecole Normale, and the Ecole Polytechnique aroused the admiration of many foreigners.[2] The British economy depended to a much greater extent than the French upon mining, and ultimately a School of Mines was set up and after 1851 moved to South Kensington where it eventually became a part of Imperial College. Also at South Kensington were the Normal School

[1] W. Coleman, *Georges Cuvier, Zoologist* (Cambridge, Mass. 1964); J. Davy, *The Life of Sir H. Davy*, 2 vols. (London 1836), II, 342ff. On the work of natural historians at the British Museum and elsewhere, see the *Historical Series* of the *Bulletin of the B.M. (Natural History)* and the *Journal* of the Society for the Bibliography of Natural History. On scientific education in the U.S.A., and ultimately on the Smithsonian Museum, see N. Reingold (ed.), *The Papers of Joseph Henry* (Washington D.C. 1972–). D. E. Allen, *The Victorian Fern Craze* (London 1969).

[2] M. P. Crosland, *The Society of Arcueil* (London 1967); A. J. Meadows, *Science and Controversy* (London 1972); C. W. J. Higson, *Sources for the History of Education* (London 1967).

where T. H. Huxley among others laboured to train and to upgrade teachers, especially teachers of science, and the Royal College of Chemistry, in which Prince Albert had taken a strong interest and which also became a part of Imperial College. The Ecole Polytechnique, with its aim of producing scientists who would be military engineers, was followed at West Point in America which seems to have given the best scientific training in America in the 1820s and 1830s, and to a lesser extent at the Royal Military Academy at Woolwich, which had a succession of eminent scientists among its lecturers but never acquired the prestige of its Parisian model. The relationship between research, teaching and social needs or demands in these various institutions cries out for further study which will illuminate the emergence of scientists as a class having common interests, for this professionalisation was what the French seemed to have achieved.

The French institutions lost momentum even during the Napoleonic period, as syllabuses became ossified, important posts were held in plurality, and cliques developed among men of science who were also, it seems, more likely than their German or English contemporaries to desert science for politics. And although French examples continued to be imitated, for example, at South Kensington, it was to Germany that attention was increasingly directed by those involved in scientific education. Philosophical and literary contacts between England and Germany in the early nineteenth century have been investigated and can be studied from books and journals published or in manuscript; but how much bearing German philosophy had upon science outside the German speaking world is open to doubt and seems imponderable.[1] German universities, on the other hand, with their

[1] H. G. Schenk, *The Mind of the European Romantics* (London 1966); F. W. J. Schelling, *On University Studies*, tr. E. S. Morgan, ed. N. Guterman (Athens, Ohio 1966); T. McFarland, *Coleridge and the Pantheist Tradition* (Oxford 1969); T. H. Levere, *Affinity and Matter* (Oxford 1971); W. R. Ward, *Victorian Oxford* (London 1965); J. Sparrow, *Mark Pattison and the Idea of a University* (Cambridge 1967); T. J. N. Hilpern, *Engineering at Cambridge* (Cambridge 1967); and the bibliographies: J. Craigie, *Scottish Education*, and H. Silver and S. J. Teague, *British Universities* (both London 1970).

emphasis on research stimulated reformers in Britain and America; by example and by actual training, particularly of chemists, they revolutionised the teaching of science and ensured that universities would become centres of scientific research. We can follow this process in the records of the various universities, in manuscripts and also in published materials. We have, for example, university calendars, which describe courses available and outline syllabuses, often in considerable detail; they often include class-lists and names of external examiners. They may also include examination papers, which may be a better guide to actual courses than published syllabuses. We can see what courses were compulsory; whether for example, Greek and Rudiments of Religion were required from all candidates. We may find, as for example at Durham in the 1830s and 1840s, that while natural philosophy was considered a proper part of a degree course, engineering led only to a kind of diploma.

But calendars and prospectuses, as everybody knows, may tell one curiously little about the functioning of a university, and certainly they are not very helpful to anyone interested in the balance of research and teaching in an institution, or even to the student of teaching methods. He may find, for example, that laboratory instruction was available on payment of an extra fee; but this will not really tell him how many students took advantage of this, or how such teaching was fitted into the general framework of the science courses. The student of science teaching in Britain at least is fortunate in that many universities were investigated by Royal Commissions, sometimes more than once; the reports of commissions are full of useful information, though little of it naturally enough may be about science. At Glasgow, for example, some professors both of sciences and arts had more power than others, and the dispute between the two groups called for outside intervention;[1] in England there was less

[1] On Lockyer, see A. J. Meadows, *Science and Controversy* (London 1972); J. B. Morrell, 'Science and Scottish University Reform', *British Journal for the History of Science*, 6 (1972), 39–56, and 'Thomas Thomson', same journal, 4 (1969), 245–65. G. W. Roderick and M. D. Stephens, *Scientific and Technical*

science taught, and the reformers often wanted more science and more research generally. One should perhaps remember that at least two presidents of the Royal Society, Humphry Davy and Benjamin Brodie, wrote that the best preparation for a career in science would be a degree in arts; and that, outside the field of applied mathematics or a medical training, was indeed the background of many eminent scientists. As in all such cases, one should not take it for granted that all right-thinking persons, or all scientists, would have been on one side rather than another in disputes about curricula and so on. As well as the inquiries into separate universities, and their federating and defederating which was a feature of the period, there was the more general report on education by the Devonshire Commission of which Norman Lockyer—later the founding editor of *Nature*—was secretary; this is a very valuable document.

Such *Reports*, taken in conjunction with official university publications and with the papers of the various institutions and of prominent members of them, can give us quite a good view of what went on; for schools and technical institutions in Britain before the Devonshire Report we have less published material and have to rely more upon manuscripts and memoirs. It is not only to the student of scientific education, naturally enough, that published *Reports* and *Parliamentary Papers* are of importance.[1] The history of technology is considerably illuminated by reports upon railways and factories, for example; and public health was during the nineteenth century a matter of great concern about which a great deal was published. The appearance of the cholera

Education in 19th-century England (Newton Abbot 1973); W. H. Brock and A. J. Meadows are writing a book on scientific education.

---

[1] P. and G. Ford, *Select List of British Parliamentary Papers, 1833–99* (Oxford 1953); *Guide to Parliamentary Papers*, 3rd ed. (Shannon 1972). M. F. Bond, *Guide to the Records of Parliament* (London 1971). D. Menhennet, *The Journal of the House of Commons* (London 1971). J. Brooke, *The Prime Ministers' Papers, 1801–1902* (London 1968). L. F. Schmeckebier and R. B. Eastin, *Government Publications and their Use* (Washington D.C. 1961); E. Jackson, *Subject Guide to Major U.S. Government Publications* (Chicago 1968).

in the 1830s, and its subsequent returns at intervals during the nineteenth century, led to the setting up of Boards of Health;[1] for example, James Kay, later Kay-Shuttleworth, summarised reports of various inspectors to give an account of the moral and physical conditions of the working classes in Manchester in 1832. Ten years later Edwin Chadwick, secretary to the Poor Law Commissioners, presented to Parliament his *Report on the Sanitary Condition of the Labouring Population* which included an immense quantity of data collected by numerous people, including Kay, and is a most valuable source of information on a whole range of topics. John Simon, who after Chadwick's fall from power was the most powerful man in the public health movement, also published reports and a history of *English Sanitary Institutions.* Industrial diseases had been investigated by Charles Thackrah of Leeds, who published a book on the subject in 1831; another surgeon, P. Gaskell, published in 1836 a gloomy account of the dangers arising from mechanisation, *Artisans and Machinery*, to set against the optimism of the chemist, Andrew Ure, whose *Philosophy of Manufactures* had appeared in the previous year. In this region the histories of science and technology impinge upon social, economic and administrative history; it is not a field for the narrow specialist.

We have already mentioned scientific expeditions and naval surveys as a major item of government expenditure on science in the nineteenth century; and governments also put time and energy into surveys of their own countries. James Rennell mapped British India in the late eighteenth century, writing a book to accompany his maps; his survey was supposed to be at least as good as anything being done in Europe. In eighteenth-century Britain, the supply of maps and charts was left to private enterprise; some European countries did better. By the

[1] N. Longmate, *King Cholera* (London 1966); J. P. Kay, *Condition of the Working Classes* (1832, reprint Manchester 1969); E. Chadwick, *Report* (1842), ed. M. W. Flinn (Edinburgh 1965); J. Snow, *On Cholera*, ed. B. W. Richardson and W. H. Frost (1936, reprint New York 1965); R. Lambert, *Sir John Simon* (London 1963).

nineteenth century large-scale triangulations were being carried on, under the auspices of the Ordnance Survey in Britain which government was prepared to finance under the illusion that mapping the country was something which could be done once and for all, rather than a continuing commitment.[1] Out of the Ordnance Survey grew the Geological Survey which again proved to be something that could not be finished in a few years; the books and reports of the directors of these surveys are valuable documents for those interested in the rise of science as a profession and in the interaction of science and government. In the U.S.A. the Coast Survey under Alexander Dallas Bache grew rapidly into the largest employer of scientists in the country; it and the Smithsonian Institution seem to have played a dominant role in nineteenth-century American science which was otherwise fragmented among small State institutions and colleges.

Naval surveys had been concerned with hydrography; that is, with charting coastline, channels and harbours. We can see from John Herschel's *Admiralty Manual of Scientific Enquiry* (1849), what kind of questions surveyors were expected to answer; the book contains sections on ethnography, economics and medical statistics, as well as on geography, astronomy and natural history. In the second half of the century the task of mapping seemed well in hand, and attention shifted to oceanography, that is, the study of the open seas.[2] Geographers like Humboldt and Rennell, and navigators such as Cook and Bougainville, had interested themselves in ocean currents; and attempts had been made to take deep soundings in mid-ocean and to study marine biology using nets and dredges. Knowledge of currents promised to shorten voyages; and Matthew Maury, Superintendent of the Depot of Charts and Instruments at Washington D.C., published sailing directions and

[1] See note 2, p. 165; C. Close, *The Ordnance Survey*, ed. J. B. Harley (Newton Abbot 1969); R. V. Tooley, C. Bricker, and G. R. Crone, *A History of Cartography* (London 1969); N. Reingold, *Science in 19th-century America*; G. E. Manwaring, *Bibliography of British Naval History* (London 1930); R. Higham (ed.), *A Guide to the Sources of British Military History* (London 1972).

[2] M. Deacon, *Scientists and the Sea* (London 1971).

then in 1853 his *Physical Geography of the Sea.* In 1872 the British government sent out the *Challenger* on a voyage around the world of more than three years, doing oceanographic research; the results were published in fifty volumes, with informal accounts written by C. W. Thomson, the scientific leader of the expedition, and H. N. Moseley, one of the natural historians, among others. The *Challenger* expedition, and the publication of its results, represented a heavy government investment in science which is still insufficiently studied.

Moseley and Thomson were professional scientists whose publications related to their science; but Moseley's father, a fellow of the Royal Society who had been professor of Astronomy and Natural Philosophy at King's College, London, published what must be one of the last *Astro-Theologies* to appear.[1] Similarly, in 1841 Parry the Arctic explorer wrote a book on the parental character of God; and as we go further back we find many men of science who wrote books on subjects which to us seem to have little connexion with what we think of as their main concern. Peter Mark Roget wrote his famous *Thesaurus*, as well as a Bridgewater Treatise on natural theology and physiology, and works on galvanism; all between 1832 and 1852. In the late seventeenth century, John Ray wrote a collection of proverbs which is a standard source, as well as his important work of natural history; and John Robison of Edinburgh has to his name both standard works on mechanical philosophy and an hysterical book, *Proofs of a Conspiracy against all the Religions and Governments of Europe*, which appeared in 1797. To read works such as these is to be reminded that academic frontiers shift as much as political ones do with time, and that scientists can have interests as wide as anybody else's. Like works of science-fiction or of popularisation by scientists, their non-scientific books may cast light on the workings of their minds even in science and may show us what 'non-scientific' disciplines were at any time closest

[1] H. Moseley, *Astro-Theology*, 2nd ed. (London 1851); for earlier ones, see the list of Boyle Lectures in J. F. Fulton, *A Bibliography of Robert Boyle*, 2nd ed. (Oxford 1961), 197ff.

to certain sciences. Thus Ray's interest in proverbs and in uncommon words went readily with a concern with natural history and taxonomy, because one of the tasks of the naturalist of the seventeenth century was to identify plants and animals described in prose and poetry.

Similarly, science and theology have always been close to one another, and authors ever since Plato's time have used material from one to cast light upon the other, though the effect has sometimes been to obscure rather than to illuminate. In our period, the relationship down to about 1850 was different in England from what it was in most European countries, in that (as in the medieval period) most men of science believed that scientific knowledge must complement Divine revelation. This doctrine sometimes led to absurdities, as when all geological phenomena were referred to the Deluge, and all variety of languages and religions to the Tower of Babel; but it gave a conviction of order, of a world which was intelligible, and it provided a link between science and ethics. At all events, from the time of the 'latitude men' of the later seventeenth century to that of the Bridgewater Treatises of the 1830s one must take natural theology in Britain seriously;[1] and during much of this period natural theology dominated theology just as natural philosophy dominated philosophy.

The historian of science therefore may be able to learn much not only from writings on natural theology specifically, which are

[1] R. E. Schofield, *Mechanism and Materialism* (Princeton 1970); A. Thackray, *Atoms and Powers* (Cambridge, Mass. 1970); D. R. Dean, 'James Hutton and his Public', *Annals of Science, 30* (1973), 89–105 and P. M. Heimann, 'Nature is a perpetual worker', *Ambix, 20* (1973), 1–25. H. R. McAdoo, *The Spirit of Anglicanism* (London 1965); D. Newsome, *Godliness and Good Learning* (London 1961), and *The Parting of Friends* (London 1966). O. Chadwick, *The Victorian Church*, 2 vols. (London 1966–70); A. O. J. Cockshutt (ed.), *Religious Controversies of the 19th century; selected documents* (London 1966). B. M. G. Reardon, *Religious Thought in the 19th century* (Cambridge 1966), and *From Coleridge to Gore* (London 1971). To see what really excited contentious theologians, see E. R. Norman, *Anti-Catholicism in Victorian England* (London 1968). T. H. Levere is writing a book on Coleridge and science.

excellent sources for those seeking the norm at any time, but also from theological treatises, lectures and sermons. A dynamical world-view, in which spirit and force were paramount over matter, is to be found in the writings of Jacob Boehme which were translated into English at the beginning of our period—a time when alchemical writings and books of natural magic were in great demand. Later came the writings of Swedenborg; and at home those of Robert Green, John Hutchinson, and William Jones of Nayland, whose piety induced them in the Renaissance manner to decode the Scriptures, and who opposed the Newtonian physics of the eighteenth century and the materialism associated with it. The historian should study the losing side as well as the winning one; and there were even eminent men of science in the nineteenth century, such as the entomologist William Kirby, who thought that Jones and Hutchinson were right. S. T. Coleridge similarly in his *Aids to Reflection*, which was a very influential work, espoused a dynamical science, derived in part from Neoplatonists and seventeenth-century authors and in part from German Romantics, and fitted the physiology of John Hunter and the chemistry of Humphry Davy into this framework. His readers would be reassured that the materialism and pantheism associated with the science of the eighteenth century had been refuted by more recent discoveries in science, and urged not to try to prove the existence of God but to make man feel the need for religion.

These authors lay outside the main stream of theological writing where the stress upon natural religion drew upon ideas of law and nature which came from the sciences; here the atomic theory, for example, was used to prove the existence of a soul distinct from matter in Cudworth, and in William Wollaston's famous *Religion of Nature delineated.* The writings of Joseph Butler are a valuable source for recovering the views of sound divines of the mid-eighteenth century; and his remarks upon natural religion and on argument from analogy cannot but interest the historian of science. In the nineteenth century, he will want to turn to those valuable collections, *Essays and Reviews* of 1860 and *Lux Mundi* of 1889; the former the production of broad churchmen and the latter that of the high-church party, but both trying to come to terms partly with modern discoveries, including those of science. He will also want to consult some of the series of Bampton Lectures and Gifford Lectures; and all this will soon convince him how much more complicated and fluid relations between science and theology were than we are sometimes tempted to suppose. Theologians, and indeed religious people generally, were probably more perturbed by Biblical criticism than by Darwinian biology;[1] and indeed agnostics were probably a minority among important scientists in Britain in the later decades of the nineteenth century. It would be interesting to know more about this, and to have comparisons between Britain and other countries.

In the German universities about the end of the eighteenth century, Theology seems to have been dethroned as Queen of the Sciences, and replaced by Philosophy. Science in Germany in the early years of the nineteenth century was intertwined with philosophy, to the horror of later scientists who saw this as an intrusion of metaphysics into what should have been empirical disciplines. There was nothing new about science and philosophy having close relations with one another; Galileo and Descartes had had a foot in both camps, and throughout our period one can find science in philosophical writings and vice versa. Locke's writings are full of explicit and oblique references to the science of his day, in which he was well informed; Berkeley's writings were often directed at specific scientific issues, or at what seemed to him to be philosophical mistakes made by scientists; and attempts have been made to trace important elements of the world-view of the Romantics in the pantheism and materialistic teleology of some

[1] G. Kitson Clark, *The Making of Victorian England* (London 1962), chap. VI; W. R. Ward, *Religion and Society in England, 1790–1850* (London 1972); S. Budd, 'The loss of faith in England, 1850–1950', *Past and Present, 36* (1967), 106–25. See the papers on the X-Club by J. V. Jensen, *British Journal for the History of Science, 5* (1970), 63–72, and R. Macleod, *Notes and Records of the Royal Society, 24* (1970), 305–22; A. O. J. Cockshut, *Truth to Life* (London 1973).

eighteenth-century biologists.[1] This last enterprise is a more difficult one; but there can be no doubt that it is usually profitable to follow up some of the references to science in the writings of philosophers, because we can thereby see interesting features both of science and of their philosophy.

Thus when in 1715 Princess Caroline came to England with her father-in-law, King George I, her former mentor, Leibniz, warned her of the dangers of English natural theology and philosophy; this challenge was taken up on behalf of Newton by Samuel Clarke who exchanged a series of letters with Leibniz in which theology, philosophy and physics are inextricably mixed up. We not only learn much about philosophy in the period from this correspondence but we also see how presuppositions in any of these three fields, which to us seem distinct, affect the conclusions to which people come in the others. In the middle of the eighteenth century there was published David Hartley's *Observations on Man* (1749), an attempt to go beyond Newton and Locke in demonstrating that all ideas could be reduced to vibrations in the brain, with a second volume to establish that such a belief would strengthen religion. At about the same time came David Hume's criticism of the procedure of induction, and of the idea of causation as implying a necessary connexion between cause and effect. His scepticism seems to have been little regarded in Scotland and England; and the philosophers, notably Thomas Reid and Dugald Stewart, who tried to refute him by 'common sense' seem to have been closely concerned with science—in which indeed induction and causality are generally taken for granted. In Germany Kant, who took Hume seriously, published works of science and derived some of his antinomies from science; and Schelling's *Naturphilosophie* was an attempt to reconstruct science in the light of Kantian philosophy. Recent studies have appeared of Goethe's science, showing how his originality lay not in his particular observations and conclusions but in his synthesis of much of the science of his day into a unified

[1] See McFarland, note 1, p. 173; and H. W. Piper, *The Active Universe* (London 1962).

view; and of Hegel's, showing how widely read he was in science, and how he, again, attempted to organise the knowledge of his day into an hierarchical structure. The close study, that is, of the science of Goethe or of Hegel can illuminate the norm of their day, and introduce us to authorities whom we might otherwise have neglected; as well as showing us more about how these exceptional and erudite men worked and thought.[1]

We have already mentioned their contemporary, Coleridge, as a theologian; he again cannot be classified simply as theologian, philosopher, critic or poet, but in all these activities he tried to bring in the science of his day. Some authors, like him and like Goethe, have tried to make science an integral part of their world-view, so that some of their writings were intended at least in part as contributions to science; others have brought in some science as a fashionable subject or to give background. Both kinds of writings are useful to the historian of science, who is interested in original syntheses and in the norm. There seem to have been few novels in our period in which the scientist is the hero; he is more likely to appear, as in Swift's *Voyage to Laputa* and in the novels of Thomas Love Peacock—both of which have been studied for the light they cast upon the history of science[2]—as a comic figure. But in essays, critical writings, and poetry we do find science being taken seriously. Thus Marjorie Nicolson has studied the impact of the telescope and microscope on the literature of the seventeenth and early eighteenth century; this kind of study not only

[1] See Schelling, note 1, p. 173; on *Naturphilosophie*, see B. S. Gower's paper in *Studies in the History and Philosophy of Science*, *3* (1973), 301–56, and H. A. M. Snelders' on Winterl and on Schweigger in *Isis*, *61* (1970), 231–40 and *62* (1971), 328–38. H. G. Alexander (ed.), *The Leibniz-Clarke Correspondence* (Manchester 1956). T. E. Jessop, *A Bibliography of David Hume and of Scottish Philosophy from Francis Hutchinson to Lord Balfour* (London 1938). H. B. Nisbet, *Goethe and the Scientific Tradition* (London 1972); *Hegel's Philosophy of Nature*, tr. M. J. Petry, 3 vols. (London 1970).

[2] M. Nicolson, *Science and Imagination* (Ithaca, New York 1956), chap. V on Laputa, others on the telescope and microscope. E. Robinson, 'T. L. Peacock; critic of scientific progress', *Annals of Science*, *10* (1954), 69–77. A. J. Meadows, *The High Firmament* (Leicester 1969).

illuminates the works of literature concerned but gives some indication of how widely-known were the discoveries of scientists.

The *Spectator* of Addison and Steele is a valuable source for this last question, as are various poems of various degrees of merit. Those favourites of the eighteenth century, the *Seasons* of James Thomson and the *Night Thoughts* of Edward Young, contain a great deal of current science—chiefly natural history in the former and astronomy in the latter poem—which is not original but is agreeably and sometimes forcefully put. The poetry of Erasmus Darwin, written by a man who had a scientific training and some reputation, does contain some original science;[1] most notably his theory of evolution. The verse, which seems to us turgid and sometimes comically ponderous in the main, was augmented by copious footnotes; its chief point was to get across in an agreeable way the idea of the sexual nature of plant reproduction, and in this it was a success. To the next generation, who sometimes used with more *panache* images from Darwin's verse, it seemed unreadable and laughable; but as a guide to the natural history, and science generally, of the 1780s and 1790s it is very valuable, and once the conventions behind it are accepted it can still be read with pleasure. The evolutionary notions in *Zoonomia* are of interest because they show the attractiveness of such ideas to the speculatively minded; conversely they indicate how much work his grandson Charles had to do to make the theory of evolution attractive to the hard-headed, even as a working hypothesis. Evolutionary ideas are also to be met with later in Tennyson's *In Memoriam*, which appeared some years before the *Origin of Species*; and in works of that curious genre, novels of religious doubt, we can see how the idea of nature as indifferent to man, and operating remorselessly on an enormous time-scale, seemed to be the consequence of discoveries in geology and thermodynamics. The historian of science, in short, must

[1] D. King-Hele (ed.), *Essential Writings of Erasmus Darwin* (London 1968); J. H. Buckley, *The Triumph of Time* (Cambridge, Mass. 1967); F. W. Newman, *Phases of Faith* (1850, reprint Leicester 1970); J. A. Froude, *The Nemesis of Faith* (1849, reprint Farnborough 1969).

read the general literature of his period and not merely the scientific writings if he is to get a full and satisfying view.

Literature is not the only fine art which has been closely involved with science. Painting has been affected not only by the chemistry of pigments but also by discoveries and theories in optics and psychology. Various scientists during our period published standard colour systems; the theory of perspective goes back to antiquity,[1] but the invention of portable versions of the *camera lucida*, for example by W. H. Wollaston in 1806, in which an image of a scene was projected on to paper so that one could readily draw round it, imposed a 'scientific' perspective upon drawings and paintings. The device was soon improved so that the image fell upon a prepared plate, and photography was born. The relation between science and aesthetics is one which can be profitably studied; not only in this connexion but for example later in the emphasis placed by Ruskin on the drawing of specimens of natural history. Music too has had a long connexion with science and was indeed long considered as a branch of mathematics; eighteenth-century travellers, for example, so describe the exotic musical traditions on which they report. To read, for example, the *Opticks* and *Harmonicks* of Smith, which were standard works of their day, is to see the intimate connexion between science on the one hand and painting and music on the other; and later on we find similar connexions in the writings of Helmholtz. Ceramics in the eighteenth century was also firmly connected to science; the attempts to reproduce the porcelain of the Chinese led to experiments in the control of composition and temperature; and the standard pyrometer of the eighteenth and early nineteenth centuries, for measuring temperatures in furnaces or estimating that of volcanoes, was that devised and described in the Royal Society's *Philosophical Transactions* by Josiah Wedgwood.[2] Wedgwood belonged, like Erasmus Darwin, to the

[1] P. Rawson, *The Appreciation of the Arts: drawing* (London 1969), 216–19; H. H. Rhys (ed.), *17th-century Science and the Arts* (Princeton 1961); V. Ronchi, *The Nature of Light* (London 1970).

[2] J. Wedgwood, in *Phil. Trans.*, *72* (1782), 305–26; *74* (1784), 358–84; *76*

Lunar Society of Birmingham; the study of this informal group casts light upon the relations between science and the fine and useful arts in the latter part of the eighteenth century. Similar studies on other groups at different period and places might be equally rewarding.

In using any books, whether literary or scientific or midway between, we must remember the general points that one always has to take into account: that, for example, there may be—especially where science is concerned—considerable changes between different editions, that material alleged to have been published in a journal earlier may have been completely rewritten, that translation may have distorted the text, or that an author may have given different titles to what were really different versions of the same book, as Joseph Glanvill was wont to do at the beginning of our period.[1] But above all we must ask who books were written for; using evidence from subscription lists, from correspondence, from reviews, from syllabuses and library holdings, and from references in other writings. In earlier periods, one can ask why books have survived; but there must be very few editions indeed published during our period which have completely perished, though many pamphlets, tracts and chapbooks will have become completely lost. We can inquire why a given book is, or is still, to be found in the library of an individual or an institution; and we can attempt to see how widely dispersed copies of a book were. The manuscripts, journals and books which we have so far discussed, and which are the most widely used sources for the history of science, do of course survive as physical objects; and more can be learnt from the original than from copies of them. But we must now turn to another series of sources; those objects, whether apparatus, machinery or specimens, which survive from the science of the past.

(1786), 380–408; vases by Wedgwood were among the presents taken on Lord Macartney's embassy to China, and are said to have been admired. R. E. Schofield, *The Lunar Society* (Oxford 1963).

---

[1] J. I. Cope, *Joseph Glanvill, Anglican Apologist* (St Louis 1956), 161–71.

CHAPTER 7

Surviving Physical Objects

In recent years our knowledge of industrial history has been greatly enlarged, because to the written sources and isolated machines preserved in museums has been added new information derived from the application of archaeological techniques to industrial sites. There is now a considerable literature of industrial archaeology, with various industries and various sites being investigated; and there are now many groups and societies carrying on the work in all kinds of places.[1] Machines were all made and processes carried on with a function, that cannot often be understood in a static display; and the newest museums of the history of technology try to show their exhibits in their context, and where possible working. While this industrial history cannot be completely separated from the history of science, its chief connexions are with economic and social history. But we can learn what materials were available; and we should always remember, for example, the long connexion between men of science and the gunpowder industry, and that the reputation of so pure a natural philosopher as Faraday was established by his investigation of the by-products of the new coal-gas industry and his researches upon glass and steels. The evidence of industrial archaeology may enable us to ask and to answer questions about the role of science in technology, where written sources are often thoroughly ambiguous. We may be able to show that the demands of industry

[1] See the journal, *Industrial Archaeology*, and the series of books on the subject published notably by Longmans, Hutchinson's, and David and Charles. See also the very valuable *Transactions* of the Newcomen Society. On chemical analysis, see F. Szabadváry, *History of Analytical Chemistry*, tr. G. Svehla (Oxford 1966), chap. VIII. J. Z. Fullmer, 'Humphry Davy and the Gunpowder Manufactory', *Annals of Science*, *20* (1964), 165–94; M. Faraday, *Experimental Researches in Chemistry and Physics* (London 1859, reprint Brussels 1969).

for a rapid means of chemical analysis, for example, led to the improvement, wider use, and eventual fuller understanding, of volumetric methods; as we find out more about what processes were used in chemical industry, in certain places and at certain dates.

But even those men of science who have most resembled lone voyagers in strange seas of thought have required some equipment; and for the abstruse reasoner as for the explorer, we shall be interested in the equipment he used and the findings he brought back. It is usually agreed that propositions which cannot be tested cannot be counted as science; and among men of science, those propositions which might in principle be tested although in practice they cannot be, are often regarded with suspicion if note is taken of them at all. Much thought and time has been devoted by men of science to the making testable in practice, by direct or indirect methods, of theoretical statements which at first seemed only in principle testable. Though we may find,[1] as with the Copernican system or the atomic theory, that a theory has become widely held because it was in accord with other beliefs widely held at the same time, we also find that attempts have always been made to confirm or refute it by experiment. It was not until the nineteenth century that the 'stellar parallax', the result of the Earth's motion predicted by the Copernicans, was detected; and only at the beginning of the twentieth century that converging lines of indirect evidence for atomism became available. To detect the stellar parallax, instruments of great accuracy were required, as well as mathematical methods of estimating experimental error and observational techniques to make such errors as small as possible. Atomism was a more complicated affair, for the various lines of inference included evidence derived from such new phenomena as cathode rays and X-rays. Theory

[1] T. S. Kuhn, *The Copernican Revolution* (Cambridge, Mass. 1967); the documents on the atomic theory are reprinted in *Classical Scientific Papers, Chemistry* (London 1968), of which I am editor. On stellar parallax, W. Hartner, *Oriens-Occidens* (Hildesheim 1968), 10, 497, 520; on cosmology, J. D. North, *The Measure of the Universe* (Oxford 1965).

and apparatus have thus reacted upon one another; and in this chapter we shall look at apparatus and other physical objects as sources for the history of science.

As well as apparatus actually used in experiments, we may find models.[1] These may have been used by somebody to clarify his thoughts, in the course of developing a theory; or in demonstrating the theory to his colleagues, or teaching it in a school or college. A model is a way of demonstrating an analogy. It may be a simple analogy, like that between a machine and a scale-model of it; or a more complex one, like that between an atom and a billiard-ball. Those using atomic models have not, it seems safe to say, supposed that their models had more than some analogies with atoms: what is interesting is to see, from the models, which properties of atoms the balls were being exploited to show. At the beginning of our period, Hooke in his *Micrographia* showed how spheres could be packed together to give the effect of some common crystal forms; and at the beginning of the nineteenth century W. H. Wollaston took this further, constructing other crystalline forms with balls which were not perfect spheres, being flattened or elongated. By the middle of the nineteenth century interest had shifted from mineralogy and crystallography towards organic chemistry; the great problem was to account in terms of arrangement of atoms for the very different properties displayed by compounds containing carbon, oxygen, nitrogen and hydrogen only. With a chemical atomism which was beginning to make testable predictions came a new kind of model in which the spherical atoms were connected together by wires; and the shape

[1] On 'models', see M. B. Hesse, *Forces and Fields* (London 1961), chap. I; and R. Harré, *Theories and Things* (London 1961). On atomic models, see my *Atoms and Elements*, 2nd ed. (London 1970); W. H. Brock (ed.), *The Atomic Debates* (Leicester 1967); C. A. Russell, *The History of Valency* (Leicester 1971); W. G. Palmer, *A History of the Concept of Valency to 1930* (Cambridge 1965). Models were also made to illustrate the relationships between chemical elements; J. W. van Spronsen, *The Periodic System* (Amsterdam 1969), and the papers in my *Classical Scientific Papers, Chemistry, 2nd series* (London 1970). O. B. Ramsay, of Eastern Michigan University, is making a study of surviving atomic models, and catalogues of them.

of the atom became of little interest, because it was the number and direction of the wires coming from the atom of each element which determined the possible arrangements. Some chemists of the 1860s criticised these models as crude and misleading; but it was very difficult to understand organic chemistry without thinking of balls and wires, and the models did help in making predictions and explaining reactions as well as in teaching the subject to beginners.

Although chemists knew that compounds were not really formed of different coloured balls or cubes held together by wires, they used these models extensively; they took it for granted because the models helped them to useful conclusions, that there were substantial analogies between the models and the compounds they represented. We cannot understand the chemistry of the later nineteenth century without thinking of these models; for the representations of formulae in two dimensions on paper was recognised as much less efficient. The close study of the various sorts of models available should amply repay the effort put into it. It has been suggested by C. A. Russell that the general acceptance of chemical atomic theory in Britain in the 1870s is connected with the rapid expansion of scientific education at that time, and the value of atomic models in teaching.

The widespread use of models has been described as characteristic of Victorian science in Britain, although in fact the ball and wire atomic models seem to have come from Germany just as much.[1] Models were used to illustrate physical properties of atoms; thus in 1884 Lord Kelvin described an 'atom' composed of rigid parts which behaved like a spring, to show that the secondary quality of elasticity could be generated from components which did not possess it. Some models of this kind may never have existed outside the head of their inventor; if they do exist as

[1] P. Duhem, *The Aim and Structure of Physical Theory*, tr. P. P. Wiener (Princeton 1954), chap. IV. Kelvin's paper is reprinted in *Classical Papers*, see note I, p. 190. J. E. McGuire, 'Forces, Powers, Aethers, and Fields', to appear in *Boston Studies in the Philosophy of Science*. J. J. Thomson, *The Motion of Vortex Rings* (London 1883).

physical objects this will tell us something about him and about his ideas. Much the same is true of the various models of the aether, the medium invoked during the nineteenth century to account for the propagation of light, and later of radio waves, and frequently also for gravitation. The aether emerged as requiring more and more extraordinary properties if it was to fulfil its various functions; and curious models, involving sometimes elaborate machinery with wheels and idle-wheels, were devised to display or account for some of these properties. Where the models were actually made, study of them in the context in which they were used may help us to decide how the aether functioned in science; what properties of ordinary matter or of energy it was invoked to account for, and how seriously its existence was believed in. William Thomson (later Lord Kelvin) at one time seriously proposed that atoms were vortex-rings in the aether, and described, in sufficient detail for us to make one, how to make a machine to produce smoke rings; the rings could then be made to collide with one another, and display other analogies with atoms or molecules which Thomson and others were at the same period exploring mathematically.

In the more descriptive sciences, there have probably been fewer models to find; but Davy, for example, made a miniature volcano containing potassium which duly erupted when he poured water into it. This model was devised to test his hypothesis of the cause of volcanic eruptions, and also to demonstrate it in a dramatic way in his lectures at the Royal Institution. Davy and his contemporaries in Britain were not familiar with real volcanoes, although Sir William Hamilton had published accounts of those in the Two Sicilies, and Cook and others had described others seen on their voyages; and when Davy did climb Vesuvius he found that the products of the eruption were not those which his theory predicted. Volcanoes are among the largest surviving physical objects with which the historian of science may have to deal,[1] bearing in mind that they may have changed their

[1] C. Daubeny, *A Description of Active and Extinct Volcanos* (London), G. P. Scrope, *Considerations on Volcanos* (London 1825). M. Rudwick, 'The Strategy

form since some particular scientist investigated them. The working of Lyell's mind has been illuminated by those who have followed his footsteps upon Mount Etna; and Darwinian pilgrims have with similar hopes visited the Galapagos Islands. To go where a great man hit upon a new idea may be simply to express filial piety; but for a descriptive scientist we may hope that if we go where he went, we shall see what he saw. It is therefore clearly desirable that anyone editing or studying a tour or voyage by a geographer, geologist or natural historian should where possible go over the ground himself.

Not only do the mountains and islands studied by past scientists still as a rule exist for us to study; but the specimens which they brought back may well also have been preserved. This is increasingly likely as we enter the more recent part of our period. Specimens were often brought back to be identified and properly described by experts at the great museums or botanical gardens; and botanists and mineralogists kept their own collections, from which they may have given the first full description.[1] The specimen thus described is the 'type' against which other specimens must be compared to see if they belong to the same

of Lyell's *Principles*', Isis, 61 (1970), 5–33. Frigid as well as igneous agencies could be revisited; see L. Agassiz, *Studies on Glaciers*, tr. and ed. A. V. Carozzi (New York 1967). See too G. L. Davies, *The Earth in Decay: a history of geomorphology, 1578–1878* (London 1969); A. Hallam, *A Revolution in the Earth Sciences* (Oxford 1973).

---

[1] F. C. Sawyer, 'A Short History of the Libraries and list of MSS. and original drawings in the British Museum (Natural History)', *Bulletin of the B.M. (Natural History), 4* (1971), 77–204. Other papers in this journal describe various collections. J. Dryander, *Catalogus Bibliothecae historico-naturalis Josephi Banks*, 5 vols. (1796–1800, reprint New York 1966). P. J. P. Whitehead, *40 Drawings of Fishes made by the artists who accompanied Captain James Cook* (London 1968); J. Ewan (ed.), *William Bartram: Botanical and Zoological Drawings, 1756–1788* (Philadelphia 1968). M. Archer, *Natural History Drawings in the India Office Library* (London 1962). On collecting, see J. Woodward, *Brief Instructions* (1696, reprint London 1973). J. E. Dandy, *The Sloane Herbarium* (London 1958); C. Waterston, *Wanderings in S. America*, ed. L. H. Matthews (London 1973).

species; when the type specimen has disappeared a contemporary drawing of it may serve instead. Much about the techniques of collecting, labelling, preserving and classifying of specimens can only be learnt from the surviving specimens. We may also learn about scientific communications, for specimens were given away, exchànged, bought and sold; they may therefore turn up in unexpected places.

In the inorganic realm there may similarly be specimens to investigate; for example in chemistry. Samples of substances prepared and analysed by J. J. Berzelius in the early nineteenth century[1] are preserved in Stockholm; elsewhere too there are in existence various nineteenth-century preparations. Examination of the specimen is always valuable where it appears that the scientist in question has made a mistake, ascribing to some compound or mineral a property which we do not find. It may be that we use the same term to describe a different substance; thus 'manganese', which to us means a metal, to eighteenth-century chemists meant the black oxide. This kind of thing can often be discovered from the literature, but is most convincingly demonstrated when we find a sample labelled by a competent man of science. Or the mistake may be a genuine one, as when Crookes' assistant supplied Lockyer's with material from the wrong bottle for an analysis. In general a modern analysis of a specimen can be very helpful in seeing what went on in an old analysis; it can also help us to estimate the efficiency of early processes of purification and the accuracy of early techniques of analysis. That is, it can cast light upon analyses which have gone wrong and those which have gone right. The difficulty is that one is as a rule confined to examination which will not use up or damage the specimen; but much can be learnt from non-destructive tests—even the appearance of the sample can tell us quite a lot—and modern methods of chemical analysis will work with very small quantities indeed. To examine the same thing, be it a mountain,

[1] J. E. Jorpes, *Berzelius*, tr. B. Steele (Stockholm 1966); W. H. Brock, 'Lockyer and the Chemists', *Ambix*, *16* (1969), 81–99, esp. 96. On analysis, see Szabadváry, note 1, p. 189.

a fossil, or a sample of a chemical compound, rather than the same kind of thing, which was described by a scientist of the past brings an immediacy which is very desirable.

The same is true of handling the equipment or apparatus which was available at the time we are interested in and is perhaps known to have been used by people with whose writings we are familiar. We may begin with descriptive science, when in geography we can clearly learn much both from surviving maps and from survey instruments, both about the end for which the maps were designed and the limitations imposed by the instruments.[1] During the eighteenth century, the accuracy of instruments for determining latitude and for making a round of angles for triangulation improved considerably, as can be seen from those which survive; the mapping of large areas on a large scale therefore became possible. At the same period, the development of chronometers made possible the accurate determination of longitude, and also the proper plotting of ocean currents; for only when a seaman knows where he is can he estimate how fast he is drifting. The chronometers themselves are some of the best examples of high precision in manufacture, though for them we have the tables of rates drawn up at observatories and on voyages which enable us to assess their accuracy without doing further trials ourselves.

With all instruments and apparatus, we have the problem of estimating how typical are the specimens which have come down to us. Most scientific equipment is worked to death and then thrown away, or used as spare parts for other pieces of apparatus; what survives may have been splendid but untypical pieces made perhaps for some nobleman, or at any rate for show rather than for use, or what happens to have been put away in a cupboard

[1] C. Close, *The Early Years of the Ordnance Survey*, ed. J. B. Harley (Newton Abbot 1969); G. S. Ritchie, *The Admiralty Chart* (London 1967); H. Quill, *John Harrison* (London 1966); *Rees's Clocks, Watches, and Chronometers* (1819–20, reprint Newton Abbot 1970); M. Deacon, *Scientists and the Sea* (London 1971); A. W. Richardson, *English Land Measuring to 1800* (Cambridge, Mass. 1966). *British Museum Catalogue of Printed Charts, Maps and Plans*, 19 vols. (London 1967).

which was not opened again for many years. We must therefore always compare the specimens we have with those described in the scientific literature and in trade catalogues; the objects and their descriptions will then complement one another to give us a fuller understanding than can be got from either source alone. This will apply particularly when we come to consider apparatus used in the laboratory.

It is with convenience and accuracy that the maker and user of survey equipment is most concerned;[1] we can look for the introduction of 'dividing engines' to draw the scale more accurately than could be done by hand, and of verniers to enable the user to read the scale more precisely. We find indeed among the different instrument makers in different countries whole series of improvements both small and great; and this is a fascinating field for those exploring the relationship between science and technology. Telescopes and microscopes were throughout our period instruments of enormous importance in science, and work has been done on the sales of telescopes and the incomes of those who made them. Here again there was a great technical innovation, the doublet lens system perfected by John Dollond in the 1750s which eliminated the coloured fringes around the image seen through earlier optical instruments. Telescopes with such lenses found a rapid sale, especially for the observations of the Transits of Venus in 1761 and 1769. Compound microscopes seem to have been improved later, possibly because the value of better telescopes was more immediately obvious; but through the nineteenth century the magnification and the clarity of the image in microscopes improved rapidly. Usually one can only look at and not through optical instruments in museums; but modern photographs have been taken of the image seen through various old

[1] On observing, see C. Babbage, *The Decline of Science* (London 1830), chap. V; S. Bradbury and G. L'E. Turner, *Historical Aspects of Microscopy* (Cambridge 1967); H. Woolf, *The Transits of Venus* (Princeton 1969); B. Land, *The Telescope Makers from Galileo to the Space Age* (London 1973); H. C. King, *History of the Telescope* (London 1955). On instrument makers, see E. G. R. Taylor, *Mathematical Practitioners*, 2 vols. (Cambridge 1954–66).

microscopes which can give us a valuable guide to what could be observed in the past.

It has been suggested that the microscope constituted a technical frontier at which men of science waited more or less patiently for opticians to provide them with an instrument which would resolve what was at present a fuzzy patch. Embryology and bacteriology may have been thus held up; and similarly in astronomy there was a long controversy about nebulae which were supposed by some to be genuine clouds of shining material and by others to be clusters of stars which no telescope had yet been able to resolve.[1] This last question was only settled with the rise of spectroscopy; that is when a new instrument and a new theory was applied to it, and the gaseous nature of some nebulae thus proved. In all sciences which have depended upon optical apparatus, there have been times when theory has been ahead of observation, so that men of science knew what to look for if only they had the apparatus; and times when observation has been ahead of theory, and data collection has been in order. This relationship in different sciences is one which repays close study; and instruments are an important source, for we need to know what performance they were capable of.

It is not only in astronomy and microscopy that one finds these frontiers. Ornithology down to the end of the nineteenth century was accompanied by what to us seems terrible carnage, because the only way to get close enough to a bird to identify it properly was to shoot it first. The development of photography and the perfection of binoculars made this unnecessary and also helped to transform the science, because bird behaviour can be readily observed through binoculars.[2] In the same way, deep sea sounding—important for oceanography—only became practicable with steam winches to haul in several miles of line, and became

[1] J. D. North, *The Measure of the Universe* (Oxford 1965), 8; W. McGucken, *19th century Spectroscopy* (Baltimore, Md. 1970).

[2] J. Huxley, *The Courtship Habits of the Great Crested Grebe* (1914, reprint London 1968): contrast H. Seebohm, *The Birds of Siberia* (London 1901), 336.

much easier at the end of the nineteenth century with the introduction of piano wire for the line.

Right through science, in short, we find this interaction between theory and the apparatus necessary to test it, which will generally also bring to light new phenomena entailing modifications to the theory. To follow this interaction we must be as familiar as possible with the apparatus. We may also observe the phenomenon of people liking to play with new toys, especially expensive ones: thus electric batteries of enormous size became a kind of status symbol, with great rivalry between London and Paris, following Davy's discovery of potassium in the first decade of the nineteenth century,[1] though the discoveries made did not measure up to the expenditure because the batteries had outrun the theories of electrochemistry. Government or other munificence in apparatus probably pays off only where there is a clear technical frontier, and good reason to think that it can be surmounted. Thus in the early nineteenth century concerted observations of terrestrial magnetism, and astronomical observations of the highest accuracy, could be made only with government support; while in chemistry the need seems to have been for thought and manipulative skill on the part of the chemist rather than for expensive apparatus or large numbers of assistants.

We have now reached the apparatus required for work in the laboratory, which probably seems to most people the paradigm of scientific activity. Here we are faced with the problem already mentioned; that ordinary apparatus has usually disappeared. An unfair sample has probably come down to us, objects made for show rather than for use surviving better than ordinary apparatus, unless for some reason the ordinary apparatus did not get used but was put safely away. The expensive items also tend to be more carefully looked after, so that balances and frictional electric machines are more likely to be found than test-tubes or retorts. When we do come across a collection of apparatus, it is clearly a help if we know its history; as we do for example in the case of the

[1] H. Hartley, *Humphry Davy* (London 1966), chap. V; M. P. Crosland, *The Society of Arcueil* (London 1967), 23–4.

chemical equipment in the Museum of the History of Science in Oxford.[1] For while such things as balances may carry the name of an instrument maker or give some direct evidence of date, glassware can be much harder to date in the absence of some evidence about the formation of the collection.

When William Ramsay took over the Chair of Chemistry at University College, London, from A. W. Williamson, he found rooms 'filled with bales, boxes, and innumerable jars and bottles' accumulated during the previous forty years.[2] The bottles and jars were unlabelled for the most part; Ramsay suggested that students should be set to analyse them, but he then worked out that this would take a hundred years, and therefore had them thrown away instead. Even unlabelled bottles would have been of some interest to the historian of chemistry today; and we must remember that such collections of apparatus as we may find will be mostly unlabelled. Scientists are taught to label their preparations, but not to put a label saying 'test-tube' on each test-tube before putting it in the cupboard; though in very tidy laboratories racks for the various pieces of apparatus may be labelled. But in general the historian will find a jumble of pieces of apparatus, and his first task will be to identify the various pieces and to work out what functions they performed and which pieces went together. His task is like that of the archaeologist identifying and piecing together the relics that he finds, carefully relating them to their context; or like that of the palaeontologist assigning the various items of his jumble of fossil bones to various kinds of creatures. The first problem then is to identify the artifact, and for this we shall require written sources, including pictures and diagrams; then from the object we can find out exactly how it worked and was handled, and thus in turn illuminate the written sources.

For identifying apparatus from the first half of our period—that is, for the later seventeenth and the eighteenth century—we

[1] C. R. Hill, *Oxford University Museum of the History of Science, Catalogue 1: Chemical Apparatus* (Oxford 1971).

[2] M. Travers, *Sir William Ramsay* (London 1956), 83.

have various sources. There are numerous works on surveying and navigation which illustrate the apparatus they describe and may thus enable us to identify it. Similarly for chemical apparatus we may turn to Rudolph Glauber's *Works* (1689), or even to the *Theatrum Chemicum* of Elias Ashmole (1652), which are handsomely illustrated. There are standard modern histories of chemical analysis, and of distillation, an operation as important for chemistry as for whisky, which anybody interested in the history of chemical apparatus and the operations for which it was used must consult.[1] For the philosophical instruments used in physical science and in surveying, there is a standard work by Maurice Daumas, of which an English translation has recently appeared; and there is a catalogue of the apparatus at Harvard in the late eighteenth century which is especially interesting since it describes a collection used in teaching at a particular time and place. There is also a very splendid work by Henri Michel on scientific instruments up to 1800, illustrated with modern photographs in which pieces of apparatus are presented as works of art, which many of those illustrated indeed were, being made for the cabinets of the curious rather than for the laboratory. But from splendidly mounted and decorated examples one can see the principles behind the workaday instruments which were made at the same period. There are also works on clocks, often devoted to particular makers, and on barometers and thermometers: time measurement was intimately connected with astronomy, the most prestigious of sciences, and the compiling of weather-reports, made suitably quantitative with measurements of pressure and

[1] Szabadváry, note 1, p. 189; R. J. Forbes, *Short History of the Art of Distillation* (Leyden 1948); M. Daumas, *Scientific Instruments of the 17th and 18th Centuries*, tr. M. Holbrook (London 1972); D. P. Wheatland, *The Apparatus of Science at Harvard, 1765–1800* (Cambridge, Mass. 1968); H. Michel, *Scientific Instruments in Art and History*, tr. R. E. W. and F. R. Maddison (London 1967). R. W. Symonds, *Thomas Tompion* (London 1951); W. E. K. Middleton, *History of the Barometer* (Baltimore, Md. 1964), *History of the Thermometer* (Baltimore, Md. 1967), *Invention of the Meteorological Instruments* (Baltimore, Md. 1969). See also C. Singer, E. J. Holmyard, A. R. Hall and T. I. Williams (ed.), *A History of Technology*, 5 vols. (Oxford 1954–58).

temperature, was a scientific activity in which anybody could join.

Of primary sources which tell us about philosophical and mathematical instruments, the standard texts of the eighteenth century are very valuable. The writings of the Dutch and English Newtonians—such men as s'Gravesande, Musschenbroek, Desaguliers, Pemberton, Hales and Emerson—are full of descriptions of apparatus for experiments and lecture demonstrations; these were, moreover, the standard works which those interested in the physical sciences read. The most famous book devoted to instruments was that of Nicholas Bion, which appeared in English in 1723; there was a second edition, with a supplement describing further instruments, in 1758, and the book is attractively illustrated. Electrical machines are described in Joseph Priestley's *History and Present State of Electricity* (1767), again with illustrations.[1] Much information about the authors of books on physical science, and about instrument-makers—and there were overlapping groups—can be found in the volumes of E. G. R. Taylor's *Mathematical Practitioners.* To see how instruments were made, we can consult Joseph Moxon's *Mechanick Exercises,* which came out in monthly parts beginning in 1678 (New Style). The first volume was devoted to mechanics generally; and to this we can turn for information about tools and about such processes as cutting teeth for gear-wheels. His second volume, on printing, will probably be of less interest to the historian of science.

In the descriptive sciences, many flower, bird and butterfly books describe not only the various species but also the equipment for catching and preserving specimens. Thus the frontispiece of Moses Harris' *Aurelian* of 1776 shows butterfly-collectors in full equipment, while F. O. Morris' *British Butterflies* (1853, and later

[1] B. S. Finn, 'Output of 18th-century electrostatic machines', *British Journal for the History of Science,* 5 (1971), 289–91; A. J. Turner, 'Mathematical Instruments and the Education of Gentlemen', *Annals of Science,* 30 (1973), 51–88. Taylor, see note 1, p. 197: on automata, A. W. G. Ord-Hume, *Clockwork Music* (London 1973). J. Moxon, *Mechanick Exercises* (1678–84), 2 vols., I reprinted (New York 1970), II, ed. H. Davis and H. Carter, 2nd ed. (London 1962); N. Bion, *Mathematical Instruments* (1758, reprinted London 1972); Priestley, note 1, p. 154.

editions) illustrated Victorian paraphernalia; George Edwards' *Essays upon Natural History* of 1770 describes how to make pictures including the feathers of birds and the wings of butterflies; and W. H. Fitton in his appendix to P. P. King's *Survey of Australia* (1827) gives pictures and measurements of geological hammers. Most standard works of the nineteenth century take these things for granted; but the valuable *Admiralty Manual of Scientific Enquiry*, edited by Sir John Herschel, which came out in 1849, enshrined the practice of half a century and is a thoroughly practical work describing how to make observations in the various descriptive sciences. It can therefore help us in identifying the equipment used.

We may have trouble in identifying an electric 'machine' or 'battery', a butterfly-net, a barrel for pickled bird-skins, or a glass case for transporting botanical specimens;[1] but our worst problems are likely to be with the smaller and humbler pieces of laboratory equipment which have become thoroughly obsolete with the introduction of pyrex glass and bunsen burners, and which were probably devised in the light of theories now quite forgotten by working scientists. A valuable source here is the technical dictionary, or encyclopedia, which indeed is valuable for the larger pieces of equipment too. Such eighteenth-century encyclopedias as those of John Harris (1704–10), and Ephraim Chambers (1728) are useful more for their definitions of terms than for their descriptions of apparatus; but we may find in

[1] On the Wardian Case for transporting botanical specimens, see D. E. Allen, *The Victorian Fern Craze* (London 1969); for the arrangements for carrying bread-fruit trees on H.M.S. *Bounty*, see W. Bligh, *A Voyage to the South Sea* (London 1792). A. M. Lysaght, *Joseph Banks in Newfoundland and Labrador, 1766* (London 1971), pt. IV; R. L. E. Collison, *Encyclopedias . . . a bibliographical guide* (New York 1964); S. P. Walsh, *Anglo-American General Encyclopedias, 1703–1967* (New York 1968). C. C. Gillispie (ed.), *A Diderot Pictorial Encyclopedia of Trades and Industries*, 2 vols. (New York 1959); C. Hutton, *Mathematical and Philosophical Dictionary*, 2 vols. (London 1795–96) should be noted; it is illustrated. Parts of Rees' *Cyclopedia* are being reprinted (Newton Abbot 1970–). D. Layton, 'Diction and Dictionaries in the Diffusion of Scientific Knowledge', *British Journal for the History of Science*, 2 (1965), 221–34.

technical dictionaries clear accounts of apparatus, and particularly of processes. Diderot's encyclopedia, and the *Encyclopedia Britannica* were handsomely illustrated, and from the plates of nineteenth-century encyclopedias we can find out much about apparatus. Particularly noteworthy for its technical descriptions and illustrations is Rees' *Cyclopedia*; this gives us a good idea of the apparatus in use across the whole of science in the second decade of the nineteenth century. General encyclopedias through the century may also answer questions like 'what was the composition of glass used in chemical apparatus in the early nineteenth century?'; at any rate, they provide an excellent place to start. Specialised dictionaries of one or more sciences are also valuable, though usually less fully illustrated; but a good dictionary of this kind should list apparatus and processes as well as substances. On the whole, such compilations are more useful when we come across a verb like 'triturate' which we do not understand, rather than when we want to know what some curious porcelain vase could have been doing in a chemical laboratory.

We may find a kind of dictionary which does answer this question. Thus Friedrich Accum, a pioneer of gas-lighting, produced in 1816 a *Practical Essay on Chemical Reagents or Tests*; this consists of a description of how to perform sundry spot-tests, and then lists those suitable for identifying various substances from 'acids' to 'zinc'. It is followed by a catalogue of chemical apparatus which Accum had for sale in his shop; this was no doubt the chief point of the publication. Such catalogues are extremely valuable because they not only describe the apparatus but tell us the cost; and trade catalogues, such as those listed in C. R. Hill's *Catalogue*[1] of the chemical collection at Oxford, are one of the best sources for the historian working with apparatus. Trade catalogues indicate when a new instrument or piece of apparatus became generally available, for there is always a time-lag between invention and first description on the one hand, and commercial

[1] Hill, see note 1, p. 200; he lists sources of information. See the proposals for a *Handbook to the Collections of Scientific Instruments in Great Britain*, *British Journal for the History of Science*, 4 (1969), 306–7.

availability on the other; and conversely, the maker's name and address may help us to date a piece of apparatus. Because firms often took on new partners, or moved, or only remained in business for a few years, a certain name and address may well fix the date within a short period of time. Where we do not find a trade catalogue, or a book by an instrument-maker which is in fact a trade catalogue, a trade card or bill-head may be valuable; for these were often elaborately engraved, and illustrate apparatus sold.

Another work which may well be by Accum is the *Explanatory Dictionary of the Apparatus and Instruments employed in the various operations of philosophical and experimental chemistry*, which was published in 1824. This useful compilation contains 17 fold-out plates and some woodcuts in the text. It is a guide to chemical manipulation, but in the context of very full descriptions, in alphabetical order, of chemical apparatus. The author remarks on the necessity for practice in laboratory operations, which cannot be learnt from books; on the way chemists were using smaller quantities than they had in the past, and thereby achieving better control over their processes; and on the close relationship between new apparatus and new discoveries. He also warns the 'theorist' that he will probably be incapable of calculating the commercial utility of any discoveries he may make, and advises circumspection. The book is very valuable for its lists of apparatus and equipment required for the laboratory and of standard reagents, as well as for its full descriptions.

Somewhat similar works were produced by the much abler chemists Lavoisier and Faraday who concentrated on describing not so much the apparatus as the processes of chemistry. In Lavoisier's case, this description forms Part III of his *Elements of Chemistry*;[1] for he hoped that all those who repeated experiments he had performed in the manner in which he had performed them would come to share his theoretical conclusions. He was a

[1] A. L. Lavoisier, *Elements of Chemistry*, tr. R. Kerr (1790, reprint New York 1965); M. Faraday, *Chemical Manipulation* (1827), a reprint has been announced, London, n.d., introduced by me.

scrupulous and accurate experimentalist: and his book therefore describes the best practice of the 1780s, in what became standard operations. Rather later, we find a nice illustration of a chemical laboratory in the 1822 edition of Samuel Parkes' *Chemical Catechism*; and a full account of how to use the apparatus in it in Faraday's *Chemical Manipulation*. Faraday was in the 1820s already at the height of his powers as a chemist. His instructions on weighing and other processes are very valuable, and his book is illustrated with wood-cuts which are useful; but Faraday was one of the greatest of experimentalists, and his book shows the wide uses to which the humblest pieces of apparatus could be put rather than the uses to which most people put it. That is, he is a poor guide to the general practice of his day, though his book is correspondingly fascinating and should make those who rely on getting all their equipment ready-made feel ashamed.

We do learn from Faraday though that chemists in the 1820s had to make their test-tubes from glass tubing, and cut out their filter-papers from large square sheets of 'bibulous paper'; small pieces of which they might use to make their own litmus-papers. We also find that chemists were beginning to make their own rubber-tubing from sheets of india-rubber; such tubes, or 'caoutchouc connectors', were a very valuable introduction because they meant that apparatus could be united with flexible but gas-tight joints. Previously, chemical apparatus had to be rigidly fixed together, the different pieces being 'luted' to one another with cements of various compositions. For strength, so as not to break under the mechanical strains involved, such glassware had to be thick; but so as not to crack under changes of temperature it had to be thin. Rubber-tubing meant that glassware could be thin, and could therefore be heated more easily. Faraday's book makes it clear that the chemist of the nineteenth century had to be a good experimentalist; an applied mathematician could be ham-fisted, but chemical theorising had to go hand in hand with laboratory work. This was a point which struck some reviewers who were able therefore to regard the chemist indulgently as a kind of advanced cook.

In the mid-nineteenth century there appeared another book on *Chemical Manipulation* by H. M. Noad (1848); this shows the standard practice of the day. There are also various monographs on particular operations or instruments, notably on the use of the blowpipe in the early nineteenth century, and on the spectroscope in the late nineteenth century, both of these being valuable new aids to the chemical analyst. And in the scientific journals one can often find descriptions, usually illustrated, of new pieces of apparatus; particularly when these had led to a new discovery. Thus Volta's first electric pile, the ancestor of our dry batteries, was described and illustrated in the *Philosophical Transactions* of the Royal Society in 1800; as was the apparatus used by Davy a few years later, in 1807, when he proved that an electric current decomposed water into oxygen and hydrogen only. Others had detected other products, which Davy attributed to impurities in the water they used, or to the effect of the atmosphere; he therefore used water distilled in silver apparatus, and electrolysed it in an atmosphere of hydrogen in vessels of gold or of agate. This tells us something about the apparatus of glass or china available in the opening years of the nineteenth century; for Davy resorted to silver and gold not because he was extravagant but because he needed apparatus that was chemically inert. The discovery, by his contemporary W. H. Wollaston, of a process for making malleable platinum meant that the chemist had at his disposal instruments and apparatus made of this extremely unreactive metal; by the time this became too expensive, because of its use in jewellery, glass good enough not to enter into chemical reactions could be had. We are unlikely to come upon large items of gold, silver or platinum in an old laboratory cupboard; but a spatula or a small crucible of platinum may have survived.

As well as journals, standard textbooks of the nineteenth century sometimes describe pieces of apparatus. This is especially likely if they are devoted to the application of a science. Thus Davy's *Agricultural Chemistry* of 1813 describes and illustrates apparatus for analysing soil; complete sets of this may be found, having therefore a greater interest than isolated pieces of equipment. When we have from one of these sources identified the objects, and perhaps the arrangement of objects, which we have come across, we can then from the objects learn more about the functions they were made to perform. We may find inventories of a laboratory at a given date; when we know what the things listed were, and were capable of, we can estimate what could and could not have been done in that laboratory. We may be able to follow improvements in some piece of apparatus which may over a period of years have become much easier to use, although its name may not have changed and therefore we would not have known about the improvements from written sources alone. Actually to see the apparatus in which familiar experiments were in their early days performed is often revealing; for we may then see that the form or material of the apparatus, or the limits of its accuracy, would have prevented earlier workers from noticing some effect which we regard as interesting. We may also find our respect increasing for those who made observations under what we would feel to be very unfavourable circumstances with clumsy equipment, in laboratories without running water, artificial light, or convenient sources of heat. We may discover the apparatus which belonged to an individual, or was used by him; thus some of John Dalton's has been described,[1] and we are thus in a better position to judge him as an experimentalist, and perhaps to revise the harsh judgements of his own and later generations. We shall be at any rate better placed in deciding to what extent accurate observations were due to native skill and industry, to luck, or to splendid equipment.

To have the artifacts is most valuable when the texts are not clear; that is, when they are discussing something which everybody took for granted, or where the author did not fully understand what he was describing. The first of their situations is not uncommon in the literature of science; the author of a book or paper will describe processes and items of equipment in terms

[1] D. S. L. Cardwell (ed.), *John Dalton and the Progress of Science* (Manchester 1968), chap. X, by K. F. Farrar. L. Trengove, 'Dalton as an Experimenter', *British Journal for the History of Science*, 4 (1969) 394–8.

which would have been immediately comprehensible to contemporaries but may mean nothing to us; he may refer for example to the lixiviation of a sample, or to an experimental set-up including Wolfe's bottle. Here the actual pieces of apparatus used, the treatises on apparatus and manipulation, and the encyclopedias and technical dictionaries of the day will together enable us to see how the processes were performed and how the apparatus worked. In the field of technology we are more likely to meet descriptions by observers who did not really understand the processes to which they were witnesses, but whose descriptions may be adequate for us to identify what they saw if we have a few examples of what it might have been. Such descriptions are a useful guide also to the possible whereabouts, or site, of some curious process about which it would be good to know more.

Those interested in the transmission of techniques, especially perhaps in the period before 1800, are likely to come across these unsatisfactory descriptions, which can only be made sense of when we can see the actual device; the written evidence is then valuable because it shows that such a device was in use at a particular place and time and may indicate whether it was taken for granted or shown off as an innovation. The work of Joseph Needham and others on the transmission of techniques has revealed how some innovations seem to have travelled readily and rapidly, while others did not,[1] this is something which demands historical explanation in the various cases. To establish that there is a genuine transmission and not a parallel development may be difficult, particularly when the form of the machine at the two remotest points is very different. Thus windmills in the East had their sails mounted to turn in the horizontal plane, while

[1] J. Needham, *Science and Civilisation in China* (Cambridge 1954–); 'The prenatal history of the steam-engine', *Newcomen Society Transactions*, *35* (1962–3), 3–58. M. Edwards, *East–West Passage* (London 1971). On steam engines, see also J. Farey, *The Steam Engine* (1829, reprint Newton Abbot 1971); the reprint is in two volumes, the second not having been previously published. See in general the Oxford *History of Technology*, note 1, p. 201.

those in the West were vertical; here therefore one would have to account for transformation as well as for transmission, and dated examples from one end of the route to the other would be very valuable evidence. The same arrangement of parts may perform a very different function in widely separated countries; thus the system of eccentric, connecting rod and piston, which is to us familiar as a means of converting reciprocating motion into circular motion—in steam engines and motor-cars, for example—seems to have originated in China where it was used to convert the circular motion of a water wheel into a reciprocating motion driving air from a kind of bellows.

This last example comes from an analysis of what Needham calls the pre-natal history of the steam engine, in which the engine is seen as an arrangement of devices which themselves originated in various places. Thus the beam of the early engines is described as a development of the Egyptian swape for raising water; the timing device, called the cataract, was a feature of Eastern water-clocks; even the metal boiler producing the steam can be given an Asian ancestry; the cylinder, valves and piston came down from Hellenistic water-pumps; and, in the later engines, the connecting-rod and eccentric, and the double-acting principle, came from Chinese pumps. The problem for such engineers as Newcomen, Smeaton and Watt was therefore to adapt and arrange these components into an efficient and thoroughly new kind of prime mover, as well as actually to manufacture such engines which were at the limits of the technology of the day. Such an analysis was not, on the whole, one which presented itself to contemporaries who saw the close connexion between the experiments on air-pressure of the seventeenth century and the working atmospheric steam engines of the early eighteenth century. But there is no reason why the historian should not be concerned with remote as well as close ancestry, though we must remember that this is a field in which hard work and close examination of sources are particularly necessary, for it is essential to provide as far as possible real evidence of genuine connexions. Conjectures about influences here as elsewhere are of little value unless backed by detailed evidence which will exclude parallel development.

For contemporary views of the steam engine and its development, we have the early nineteenth-century writings of John Farey and Thomas Tredgold (1827), which are of value both for the history of the engine and for the evidence they provide of the method of construction, cost and efficiency of various particular engines. Similarly for coal-mining we have the writings of Robert Galloway;[1] and various earlier illustrated works such as that of T. H. Hair, which enables us to identify tools used in the mines. For engineering generally, there is Edward Cresy's useful *Encyclopedia* of 1847. But the detailed history of technology is a sphere into which we cannot here enter; the historian of science who requires a guide to sources there should consult the *Oxford History of Technology*, Ferguson's *Bibliography* of technology, the catalogues of museums, and recent works on industrial archaeology. It was with industrial archaeology that we began this chapter, and we have therefore come full-circle. Since science is a practical activity, in which specimens and instruments have a considerable importance, surviving specimens and instruments are sources which constitute a valuable supplement to the written sources, which in many cases cannot be fully understood without them. They also provide evidence in cases where there are no written sources at all, though here they must be handled with all the caution appropriate to archaeological finds. With these surviving physical objects, we come to the end of our list of the various kinds of sources available to the historian of the science of our period, though no doubt future historians will think of using for

[1] R. Galloway, *A History of Coal Mining in Great Britain* (1882, reprint Newton Abbot 1969), ed. B. F. Duckham. T. H. Hair, *Views of the Collieries of Northumberland and Durham* (London 1844); on illustrated works on technology, see F. Klingender, *Art and the Industrial Revolution*, ed. A. Elton (London 1968). E. S. Ferguson, *A Bibliography of the History of Technology* (Cambridge, Mass. 1968). A. Thackray, 'Science and Technology in the Industrial Revolution', *History of Science*, *9* (1970), 76–89. C. S. Smith, *Sources for the History of the Science of Steel, 1532–1736* (Cambridge, Mass. 1968).

evidence materials which we have overlooked. The time has therefore come to return to the general problem with which we set out: what sort of questions can and cannot be sensibly asked and answered, given the sources which we have?

Epilogue

The great W. H. Wollaston, whom we have often met before, was so keen-sighted that it was said that he could botanise on horseback. But he has not left behind him a reputation as a botanist; for that science was chiefly advanced by those who grovelled about on mountain sides, peered through microscopes, and carefully described and classified specimens preserved in great herbaria. But in botany both the overall view of the flora of a region, and the detailed descriptions of species, are valuable. The general view must be controlled by detailed knowledge, and the detailed descriptions given point and meaning by general views. So in the history of science, we must know when to proceed on horseback and when to go on hands and knees; and it would be deplorable if specialisation were to lead to a situation where some historians collected the facts and others generalised from them, as in Bacon's *New Atlantis*.

In the same way, the history of science itself is not an enclave into which properly qualified experts alone may enter. It is a field for which many of the sources are common to other branches of history, and the problems similar. Economic, social, ecclesiastical, military and political historians will frequently or on occasion need to interpret the same manuscripts, journals, books or artifacts as historians of science. Because science is an intellectual, a social, and a practical activity, its ramifications stretch out into all sorts of other fields, and everybody interested in history can and should be interested in the history of science; the frontiers which separate it from other branches of history are artificial and should be disregarded unless they are in a given situation convenient. Similarly the historian of science must not specialise too narrowly, but must try to cover a wide range; this will mean that he should not restrict himself in general to one science, but will need to restrict himself in time. Histories of chemistry, for example, from the neolithic period to the discovery of DNA should be looked at with the same suspicion as any other kind of history covering such a span; though they may be illuminating they should not be taken as the norm, and a great many questions upon which the sources may cast light cannot be raised in such a context.

Science, we have said, is an intellectual, social and practical activity. It is upon the first of these aspects that most of those whom we think of as historians of science have perhaps concentrated. Here the sources are the various manuscripts, journals and books, though we may need to look at instruments, or at least descriptions of them, as well. There is indeed no class of sources which illuminate only one aspect of science. Manuscript and printed materials may help us to discover who influenced whom, and who discovered what, though, as we saw earlier, these may be unprofitable questions. We use the same kind of sources, but particularly the printed ones, in trying to determine what counted as a scientific explanation at a given period. The sciences have never been monolithic, and at any time various kinds of explanations have come into favour. In the seventeenth century, teleological explanation, despite its use by Harvey, fell out of favour and began to seem unscientific, as we can see in the writings of scientists and philosophers, and the manifestos of scientific societies. It was replaced by mechanical explanation and by mathematical explanation; the tension between these two can be seen, for example, in the reception of Newton's work, for he seemed to have provided the equations which accounted for and predicted planetary motions although no intelligible mechanical picture could be formed of how the sun acted at a distance upon the planets. This tension is one which has persisted into the science of our century and is therefore to be found in various sciences right through our period. The nineteenth century saw a resuscitation of teleology in, for example, the biology of Cuvier, and then its absorption into a version of historical explanation in the theory of evolution. This also absorbed the view that to classify is to explain, a view characteristic of eighteenth-century natural historians, which has been again revived in our century by physicists working on fundamental particles. They have also extensively employed statistical explanations which entered physics in 1859 with Maxwell's theory of gases and have been prominent though not always fully acceptable ever since.

We can hope to discover, by consulting writings on science and on the philosophy of science, what kind of explanation was preferred at a given time, and to see how changes have come about within sciences as different canons of explanation have come into fashion, necessitating different theories and new observations and experiments. Science as an intellectual activity is connected with other intellectual activities: another line we can pursue is to follow the relations between the various sciences, and between sciences and literature, philosophy and theology. The sources here will again be predominantly written, with reviews and general journals and popular writings being more important than specialised journals and monographs, though even these cannot be ruled out. It would be interesting to know whether some poet or bishop owned some scientific apparatus, as for example Shelley and J. H. Newman did during their undergraduate careers. We might see whether works of science provided images for works of literature—this being a field where it is wise to proceed with caution—and, more seriously, how methods of proceeding and notions of explanation were at different times in our period transferred from prestigious sciences such as astronomy, chemistry and geology into less-favoured sciences and across into other branches of knowledge. When we can fix our norms of scientific explanation and scientific method at different times—remembering that at any time there will be various competing norms—we can hope to see how science has influenced other disciplines. This is a process for which the documentary sources exist; it requires the historian to take a broad view of the sciences in any period and will help him to see what 'science' meant to contemporaries and where they drew their frontiers between various disciplines.

This brings us to science as a social activity; for in our period scientists emerged as a self-conscious group. This was in part connected with the growth of scientific societies and academies, and in part with the beginning, in the nineteenth century, of formal courses at universities or elsewhere for scientists. In the nineteenth century, again, we find an increasing number of jobs available only to qualified scientists. To determine which are causes and which effects in such general processes is probably hopeless; but the sources can help us to study the appearance of the scientific community, and the professionalisation of science. The pursuit of science as a career became possible in the nineteenth century in a way that it had not been earlier, when the great majority of men of science had had independent means or some profession. This process we can study from the biographies of scientists; from the papers of companies which employed them; and from the records of universities and colleges which trained them, and in turn provided lectureships and professorships, and laboratories for teaching and for research. The papers of those who played a prominent part in such institutions are a valuable source, often telling us more than more formal records, printed or in manuscript, do; and any surviving apparatus used in industry, in teaching, or in research, tells us much about what went on in these various laboratories.

Universities and colleges aroused government interest and increasingly began to need government support; and industries and communications came under increasing government regulation. Whether we study the early years of railways[1] as a development of science and technology; in connexion with economic growth; or to cast light on the development of the civil service; we shall need to use Parliamentary Papers as well as newspapers, printed books and illustrations, and the papers of those concerned in the enterprise; and naturally here again surviving track,

[1] See e.g. J. Simmons, *The Railways of Britain*, 2nd ed. (London 1968), and his essay in H. J. Dyos and M. Wolff (ed.), *The Victorian City*, 2 vols. (London 1973), I, 277–310. H. W. Parris, *Government and the Railways in 19th-century Britain* (London 1965).

locomotives or carriages are a very valuable source of information. Earlier than this, but continuing right through our period, there had been government support for and interest in voyages of exploration, and in astronomy which promised the key to better navigation. The Royal Society in Britain, and Academies of Science in other countries, were engaged in frequent negotiations in connexion with expeditions, and the reports of the various committees concerned are a valuable and sometimes fascinating source. The final report is often printed at the beginning of a voyage or is incorporated in the official instructions which are printed there; but light can often be cast on the report by study of the papers of those involved. Agriculture and public health were other fields in which government encouraged and took note of the views of men of science individually, and of scientific societies. In all these fields, there will be materials in all our categories; particularly valuable will be the papers of those men of science who went on to play a prominent part in public life, such as, in the nineteenth century, Lyon Playfair or John Lubbock in Britain, and J. B. Dumas and Marcelin Berthelot in France; for men with a foot in both camps occupied a vital position in the interaction between science and government.

We also find right through our period government interest in science as a key to improving armaments. Chemists were involved in gunpowder manufacture, natural philosophers investigated the paths of projectiles and the recoil of guns, and explorers brought back reports on the military condition of distant countries as well as on their fauna, flora and geography. Governments, in short, seem to have promoted science partly for reasons of prestige, but chiefly because it was practical. There is much still to be learned about the relationship of sciences to technology in the eighteenth and nineteenth centuries, a relationship that was probably very different at different periods and in different places. Whether science was or was not closely connected with major technological innovations is one question; whether it was thought to be closely connected is another, and to answer this question we must look at the writings of the numerous publicists and popularisers of science, at those who gave evidence at inquiries into industries or universities, and at those who delivered public lectures or presidential addresses to societies for the advancement of science. Here again then we shall find a mixture of written sources, published and in manuscript, and some surviving objects, such as the steels and glasses made by Faraday at the Royal Institution.

If science was a practical activity in that it had connexions with technology, it was also practical in that much of it was done in the laboratory. As we saw, it is only when we know what apparatus was available that we can determine whether or not a given scientific proposition could have been verified or falsified. In science a keen sense of what will or will not be a convincing experiment, and of what is in detail rather than in principle verifiable, is a valuable gift indeed; and to know the limits of accuracy at a period, and to know also how aware men of science were of the limits, is vital for the historian. In this field the apparatus itself, is, when it survives, the most valuable source of information; but it cannot be understood without some help from printed and manuscript sources. In many cases too the apparatus will not exist; but there may be plenty of information about its performance in laboratory notebooks, field notes, or published accounts of the work for which it was used.

While therefore some kinds of sources are more important in some fields of inquiry than are others, there is no field in which we can exclude any kind of source; we shall always have to find our way into the field with background reading, then come to grips with primary published sources, in books and journals, and then look at manuscripts and at surviving objects. The actual sources we use—the specific books, journals, letters and objects—will naturally depend upon our field of inquiry, as will the use that we make of them. The historian of technology and of astronomy will for example probably examine a telescope differently; and the historian of botany will not use Joseph Banks' correspondence exactly as the student of the social history of science will. To the astronomer, the telescope was a means, while to its maker it was an end; and Banks was a natural historian whose abilities and social position gave him a place of great power in the world of science; contemporaries and historians see people, events and objects in more than one light.

If science is then an intellectual, social and practical activity, ramifying out into all sorts of other fields, it would be misleading to write its history as though it were the heaping up of indubitable facts. Believers in progress have derived a comfort from the history of science which they cannot find in other branches of history; whether or not there is real progress, there can be little doubt that science is a dynamic activity, a process in which argument is a sign of health and in which the questions asked are at least as important as the answers given; and this is a view that the sources seem to confirm. It is a pity therefore that much written in the history of science is dull and pedantic;[1] in the attempt to achieve 'maturity' too many historians of science have followed the path of 'normal science' in looking at trees rather than woods. History of science in which general questions are lost sight of, or in which it is supposed that events exhaustively described somehow explain themselves, is untrue to its subject; for in both science and in the history of science events need to be evaluated and interpreted.

It would be well if the historian of science were to concentrate upon recovering the norm in science at a given time; that is, on seeing what 'science' involved, as well as what facts were known to experts or to laymen; and on investigating the relations of science to other activities. In these regions there are abundant and valuable sources, many of them hitherto unused so that the historian can feel the pleasure of discovery as well as of reinterpretation. On the other hand, it is probably best to avoid searching for hitherto neglected geniuses who anticipated well-known discoveries, and indeed to avoid arguments about priority altogether unless contemporaries were involved in them and we therefore have sources to guide us. Similarly, it is wise to eschew

[1] J. Agassi, 'Towards an historiography of science', *History and Theory*, Beiheft 2 ('S-Gravenhage, 1963).

the application of twentieth-century labels like 'operationalism' or made-up labels like 'progressionism', to the science of the past;[1] for here the sources can give us no information, and we shall both mistake labelling for understanding and find ourselves lumping together men and doctrines not grouped by contemporaries. Unless we have good evidence from sources, we shall also be careful of writing too much about influences; the sources will reveal sufficient numbers of connexions, and without good evidence we must not add to the number. Above all, the history of science is an attractive and important field of history, in which the sources are abundant and varied, and rich harvests are to be gathered; men of science have been very varied in their interests and background, and in the proportion of their time which they have devoted to science; historians of science can follow this example, for theirs, like any other activity, is not one which can be profitably confined to experts.

[1] 'Positivism' is perhaps unwisely applied as a label in my *Atoms and Elements*, 2nd ed. (London 1970), because those described were not disciples of Comte. For curious labels, see E. Meyr's introduction to the reprint of C. Darwin, *Origin of Species*, 1859 (Cambridge, Mass. 1964), and L. Eiseley, *Darwin's Century* (London 1959).

Index